Relatively
Speaking

W · W · NORTON & COMPANY

NEW YORK · LONDON

ERIC CHAISSON

Space Telescope Science Institute

Relatively Speaking

Relativity, Black Holes, and the Fate of the Universe

Illustrated by Lola Judith Chaisson

PHOTO CREDITS: *The author is pleased to acknowledge the following sources for many of the illustrations reproduced in this book.*

Frontispieces for chapters 1 and 16 and figures 2 and 53a and b, National Optical Astronomy Observatories; chapter 3 frontispiece, Meggers Collection, American Institute of Physics Niels Bohr Library; chapter 14 frontispiece and several chapter 9 photos (with Elsa, sailing, with Planck, with Michelson and Millikan, with Menzel and Birkhoff, with Shapley [Shapley Collection]), AIP Bohr Library; frontispieces for chapters 4 and 10 and figure 8 and chapter 9 figure (alone in Berlin), Einstein Archives, The Hebrew University of Jerusalem; chapter 7 frontispiece, United States Air Force; figure 35, Bell Laboratories; figure 19, CERN; figure 22, Lick Observatory; chapter 15 frontispiece, John Bedke, Carnegie Institute of Washington; chapter 17 frontispiece, John A. Wheeler (photograph by W. Litwin and J. Krumgold); chapter 19 frontispiece, NASA; figures 46, 47, 49a, and 53c, Harvard-Smithsonian Center for Astrophysics; figure 49b, MIT Haystack Observatory; chapter 9 frontispiece, Philippe Halsman; epilogue frontispiece, chapter 9 figure (with Dukas), Lucien Aigner; chapter 9 figure (on bike), California Institute of Technology; chapter 9 figure (at blackboard), Brown Brothers; chapter 8 frontispiece, chapter 9 figure (with Chaplin), AP World Wide Photos; chapter 9 figures (in mid-1950s and of empty office), Alan W. Richards.

The text of this book is composed in Avanta, with display type set in Typositor Centaur. Composition and manufacturing by the Haddon Craftsmen Inc. Book design by Marjorie J. Flock.

First published as a Norton paperback 1990

Library of Congress Cataloging-in-Publication Data

Chaisson, Eric.
  Relatively speaking.

  1.Relativity (Physics) 2.Cosmology. 3.Black holes (Astronomy) I. Title.
QC173.55.C46   1987   530.1'1   87–11134

ISBN 0-393-30675-5

W. W. Norton & Company, Inc., 500 Fifth Avenue, New York, NY 10110
W. W. Norton & Company Ltd., 10 Coptic Street, London WC1A 1PU

  4 5 6 7 8 9 0

To the late editor of the Harvard College Observatory,
LYLE GIFFORD BOYD,
who impressed upon me the need to disseminate science

Contents

Preface

RELATIVITY, the quintessence of modern scientific theory, normally evokes sighs and even fears from nonscientists. Albert Einstein's creation is often judged to be the purview only of the geniuses of our society. He himself admitted that "the non-mathematician is seized by a mysterious shuddering when he hears of 'four-dimensional' things." Surprisingly, however, those nonscientists who give the subject some careful thought succeed remarkably well in achieving at least a broad appreciation for relativity's central concepts and implications. I suggest that anyone willing to forgo common sense and human intuition can grasp the essentials of this, the grandest accomplishment of the physical sciences.

This book is based upon a series of lectures delivered during the past decade to nontechnical audiences at Harvard University. As in the lectures (parts of which are published elsewhere),* I aim

*"Black Holes, the Fate of the Universe, and Other Matters," in *After Einstein,* Peter Barker and Cecil G. Shugart, eds. (Memphis: Memphis State University Press, 1981), and *La Relatività* (Milan: Gruppo Editoriale Fabbri, 1983).

here to share the fundamentals of relativity theory, as well as some of its cosmological and black-hole applications, with anyone having an abiding curiosity about the nature of our world.

ERIC J. CHAISSON

Harvard, Massachusetts
Autumn 1986

. . . I enter the black hole. Seeing the shadow at my feet lose itself in the darkness, I have the impression of plunging into icy water. Before me, at the very end, through the layers of black, I can make out a pinkish pallor. . . . I stop to listen. I am cold, my ears hurt, they must be all red, but I no longer feel myself; I am won over by the purity surrounding me; nothing is alive, the wind whistles, the straight lines flee in the night. . . .

——Jean-Paul Sartre

INSIGNIFICANCE

Prologue
Broadest View of the Biggest Picture

[T]he cosmic religious experience is the strongest and the noblest driving force behind scientific research. No one who does not appreciate the terrific exertions, and, above all, the devotion without which pioneer creations in scientific thought cannot come into being, can judge the strength of the feeling out of which alone such work, turned away as it is from immediate practical life, can grow. What a deep faith in the rationality of the structure of the world and what a longing to understand even a small glimpse of the reason revealed in the world there must have been in Kepler and Newton. . . .

—A. Einstein

HUMANS HAVE SOUGHT, probably since the dawn of civilization, to understand the nature of our world. They thought about themselves and their environment, their planet and the cosmos.

Only a few hundred years ago, however, did natural philosophers become aware that thinking about nature is not enough to advance a viable concept. Renaissance workers declared that looking at nature is also useful, indeed necessary. Experimental philosophy, now called observational research, thus became a central part of the process of inquiry. To be effective, concepts must be tested experimentally, either to refine them if experiment favors them or to reject them if it does not. The scientific method was born—the most powerful machine ever conceived for the advancement of verifiable information.

Interestingly enough, application of the scientific method, especially in the past few decades, has demonstrated that we living creatures inhabit no very special place in the Universe at all. It is a sobering thought that we live on what seems to be no more than an ordinary rock called Earth, one planet orbiting an average star labeled the Sun, one star system in the suburbs of a much larger swarm of stars termed the Milky Way, one galaxy among countless billions of others spread throughout the observable abyss called the Universe.

Beyond our Milky Way reside uncounted billions of galaxies —enormous assemblages of matter, each containing about a hundred billion stars. Light and other forms of radiation travel outward in all directions from these galaxies, carrying the message of their existence. Only a minute portion of that light is intercepted at Earth. And because light does not travel infinitely fast—it travels at a finite velocity, the velocity of light—even light needs time to move through the unimaginably vast realms separating objects in the Universe.

Astronomers, then, are historians; our telescopes, effectively time machines. Looking out from Earth, we see a history of the Universe before us. Scanning into space is equivalent to probing back into time. Much like archaeologists who dig for hints and clues contained within old artifacts and fossilized bones, astronomers sift through ancient radiation only now arriving at the planet Earth. Yet astronomers study more than the origins of men and women. By looking out far enough, with the best telescopes, astronomers can address the origins of matter and (maybe) energy itself.

Today's astrophysicists stand at the threshold of describing the biggest picture of all—the grand design of the complete Universe. Our subject is cosmology—the study of the origin, evolution, and destiny of the gargantuan aggregate of matter and energy composing the cosmos in its entirety.

For purposes of understanding the bulk properties of the whole Universe, the smaller objects such as planets and stars— even galaxies, to a certain extent—become nearly irrelevant. To the cosmologist, planets are of negligible importance, stars hardly more than point sources of hydrogen consumption, and galaxies mere details in the much broader context of all space. Likewise, time pales in significance when compared with eternity. An interval of a million years is but a wink of an eye in the cosmic scheme of things. Even a billion years encompass a rather brief duration in the context of all time. To appreciate cosmology, we must broaden our view to include all of space and all of time. Tritely stated though no less true, to understand what follows in this book, each of us must strive to think big!

Modern cosmologists employ two primary tools. One, a set of facts, is the combined motion of all the galaxies. The other, a theory, is the concept of relativity. Observations of galaxies are the province of astronomers. Relativity theory, until rather recently, was the province of Albert Einstein. In subsequent chapters, we consider each of these tools in turn.

A BRIEF PERSPECTIVE

A more complete system of theoretical physics is made up of concepts, fundamental laws which are supposed to be valid for these concepts and conclusions to be reached by logical deduction. It is these conclusions which must correspond with our separate experiences. . . .

The structure of the system is the work of reason; the empirical contents and their mutual relations must find their representation in the conclusions of the theory. In the possibility of such a representation lie the sole value and justification of the whole system, and especially of the concepts and fundamental principles which underlie it.

These fundamental concepts and postulates, which cannot be further reduced logically, form the essential part of a theory, which reason cannot touch. It is the grand object of all theory to make these irreduceable elements as simple and as few in number as possible, without having to renounce the adequate representation of any empirical content whatever. . . .

—A. Einstein, *Essays in Science*

I

Remote Galaxies

OUR CIVILIZATION may never develop the technology to journey far enough from the Milky Way to look back and see its full grandeur. The Milky Way galaxy seems just too big to get beyond. In total, it spans more than a hundred thousand light-years, or a billion billion kilometers. (A light-year is the *distance* light travels in a single year.) Yet, from our vantage point at Earth, we can study the variety and distribution of other galaxies—colossal star systems far beyond our own Milky Way. To do so, we must use large telescopes to observe the skies perpendicular to the plane of the Milky Way, thereby avoiding the obscuration caused by the dust inherent to our own Galaxy.

Figure 2 illustrates a region of space containing myriad objects looking strangely unlike stars. Immanuel Kant, the eighteenth-century German philosopher, regarded each of these fuzzy, lens-shaped objects as "island universes" well beyond the confines of our Milky Way Galaxy. To label each of them a "universe"

The Andromeda Galaxy

FIGURE 1 An artist's conception of the Milky Way Galaxy, airbrushed on the basis of many studies of (mostly) radio and infrared data accumulated during the past two decades. The spiral arms of stars as well as loose interstellar gas and dust extend in a disk roughly 100,000 light-years beyond the Galaxy's center. The arrow within the penultimate spiral arm (toward the top of this drawing) is meant to represent the location of our Sun, a rather undistinguished star some 30,000 light-years from the center. The two blotches to the left are the Magellanic Clouds, seen only from the Southern Hemisphere, and airbrushed here to proper size and scale from the perspective of an observer roughly 300,000 light-years above the disk of the Galaxy. Recent observations imply that invisible gas extends in the form of a galactic halo or corona well beyond the delineated spiral arms; such extraordinarily thin, cool gas might well fill the entire frame of this illustration, thereby making our Galaxy much larger than heretofore envisioned and possibly engulfing the Magellanic Clouds within the Milky Way's invisible halo.

presents an obvious semantic problem, but he was correct in arguing that these nonstellar patches of light are far beyond our Galaxy. We now recognize each of these remote beacons—in fact, nearly every point of light in figure 2—as a separate galaxy. Each is a vast collection of matter comparable to our Milky Way, measuring hundreds of thousands of light-years across. And each is a gravitationally bound assemblage housing hundreds of billions of stars, many probably having planets, some possibly harboring intelligent life. Furthermore, most have spiral arms composed of stars, gas, dust, cosmic rays, and radiation much like those in our own Galaxy. Silently and majestically, each galaxy twirls in the deep reaches of the Universe, granting us a feeling both for the immensity of the Universe and for the minuteness of our position in it.

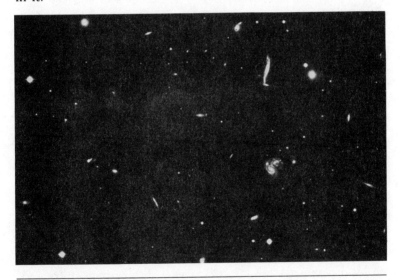

FIGURE 2 A cluster of galaxies, each galaxy of which harbors hundreds of billions of stars—more stars than people who have ever lived on Earth. This cluster is named Hercules and resides some half-billion light-years away. Galaxy clusters are among the largest assemblages in the hierarchy of material coagulations in the Universe; they probably contain a great deal of hidden matter in the intervening dark spaces among the member galaxies of each cluster.

Many galaxies are tightly grouped in so-called galaxy clusters. Each galaxy within a cluster has some random motion resembling the chaotic activity of molecules in a parcel of hot air or the turbulent motions of liquid in a pot of boiling water. On the largest-possible scale, we might then expect the galaxy clusters to have a random, disordered motion—some moving one way in the Universe and others a different way. But this is not so. On the grandest scale of all, the galaxy clusters have a definite, organized movement. And this movement is recessional: outside the local realm (that is, beyond a few tens of millions of light-years), all the galaxies seem to be moving away from us. An apt analogy might be a jar full of fireflies that has been heaved away; the fireflies within the jar have random motions because of their individual whims, but the jar as a whole, like a galaxy cluster, has a directed motion as well.

How do we know that galaxies share this net, directed motion away from us? The answer is that the galaxies' spectral features are red shifted. Such emission and absorption features arise from the galaxies' numerous chemical elements and are indicative of the velocities of those galaxies. The reddening of these features (or lengthening of their wavelengths) implies that the galaxies are steadily receding; it is as though the galaxies' emitted light waves were stretched owing to their retreating motions. This is a Doppler effect, much as horns on trains speeding away sound lower in pitch. Furthermore, the extent of the red shift increases with increasing distance to each object, suggesting a connection between Doppler shift and distance. This trend of greater red shift for objects farther away holds valid for virtually all galaxies beyond the local realm. (A few nearby galaxies display blue shifts and hence have some motion toward us, but this results from their random motions within the nearest galaxy clusters.)

Not only are the galaxies receding, then, but they are also receding at velocities proportional to their distances. A linear relationship—a perfect correlation—connects velocity and distance. The greater the distance an object is from us, the faster it recedes.

Figure 3 shows a diagram of recessional velocity plotted against distance for numerous galaxies within four billion light-years of our vantage point here on Earth. Plots like these were first made by the American astronomer Edwin Hubble in the late 1920s and hence bear his name; the resulting statistical analysis, shown by the solid line, is sometimes called Hubble's relation.

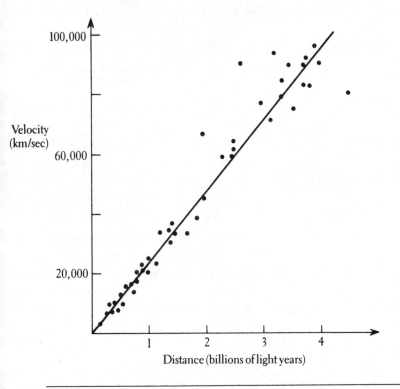

FIGURE 3 This is a Hubble diagram, a plot of recessional velocity against distance for many galaxies within four billion light-years of Earth. The solid line constitutes the "best fit" to the observed data points and suggests that the farther away the galaxy, the greater its velocity. The result is Hubble's law, considered by virtually all contemporary astronomers to be indicative of the rate at which the Universe is expanding.

Such a diagram can be made for any group of galaxies for which the distance and the radial velocity have been measured.

As can be noted directly from figure 3, galaxies at the remote distance of four billion light-years speed away with velocities of roughly 90,000 kilometers/second. This is a fair fraction of the velocity of light, the Universe's ultimate speed limit of 300,000 kilometers/second (which, by the way, is a rounded-off value meant to approximate actual light velocity in a vacuum, 299,793 ± 1 kilometers/second). The galaxies' velocity is considered genuine, though; the red shift is actually measured. The ratio 90,-000/300,000 means that such a distant galaxy has a recessional velocity of some 30 percent of the velocity of light. That's fast, very fast.

Hubble's relation is an empirical discovery, based strictly on observational results. Its central relationship—a statistical correlation between velocity and distance—is known to be valid out to at least four billion light-years. But there is no basic physical reason for the existence of such a relation. No law of physics demands that all galaxies recede. And no physical principle requires distance and velocity to be so correlated. Accordingly, astronomers are currently unsure if this relation holds true for astronomical objects much beyond several billion light-years.

Don't be confused here. Good and basic physical reasons suggest that red shift is an indicator of recessional velocity. This is the Doppler effect; the larger the spectral shift, the greater the net motion between the observer and the observee. But the Doppler effect in no way relates velocity and distance. In particular, it does not predict Hubble's relation at all. Hubble's relation is strictly a compact way of noting the observational fact that any galaxy's recessional velocity seems directly related to its distance from us.

By quantifying Hubble's relation, we can make it more useful. This can easily be done since velocity and distance are linearly related. The solid-line fit to the data of figure 3 can be expressed by a simple proportionality:

recessional velocity $\propto$ distance

Or, without the proportionality sign, we could write the equation:

recessional velocity = Hubble's constant $\times$ distance

Here, the proportionality factor between velocity and distance is called Hubble's constant. We can derive its value by estimating the slope of the best-fit line in figure 3—about 90,000 kilometers/second divided by 4 billion light-years, or approximately 22 kilometers/second/million light-years. Thus, for every additional million light-years of distance from us, astronomical objects race away with an added speed of some 22 kilometers/second.

This is the best current value for Hubble's constant, though it is a statistical approximation to the data plotted in the Hubble diagram. Over the years, new methods (and better calibration of older methods) used to determine distance have repeatedly forced astronomers to revise the value of Hubble's constant; fifty years ago, it was thought to be some ten times larger. Even today, astronomers suspect that the value of the Hubble constant might be skewed a bit by the drift of the local cluster of galaxies (containing our Milky Way) toward a much larger group of galaxies, called the Virgo Cluster; this net drift amounts to about 600 kilometers/second and might be nothing more than the random motion of our Galaxy in the outskirts of a gargantuan "supercluster" of galaxies. At any rate, modern researchers regard the current value to be accurate to at least within a factor of two; thus, it might be as much as double or as little as half the above value, but we do not anticipate the need for further major revision.

Throughout the past few decades, astronomers have strived to refine the accuracy of Hubble's diagram and the resulting estimate of Hubble's constant, for this constant seems to be one of the most fundamental quantities of nature. It specifies the rate of movement of the grandest contents of the Universe. As such, Hubble's constant is a cornerstone of any study of the large-scale structure, origin, and destiny of the Universe in toto.

For now, suffice it to note that Hubble's relation implies that all the galaxies emanate from a point, perhaps from the site of an explosion at some time in the remote past. (Admittedly, it seems that we are at this point—at the center of the Universe—but in the next chapters I shall relate how relativity theory maintains that this is incorrect.) The more distant an object is from us, the greater the force with which it must have been initially expelled; the faster-moving galaxies are by now farther away *because* of their high velocities. Bomb fragments form much the same pattern in the aftermath of an explosion.

The recessional motions of the galaxies prove that the whole Universe itself is in motion. Not really a quiescent pillar of stability, the Universe is changing with time—in short, evolving. Not only does the cosmos expand, but it also does so in a nonrandom, directed fashion. Be assured, though: neither Earth, nor our Solar System, nor the galaxies themselves are physically ballooning in size. These coagulations of rocks, planets, and stars are gravitationally bound and are not expanding. Only the largest framework of the Universe—the ever-increasing distances separating the galaxies and especially the galaxy clusters—manifests this cosmic expansion.

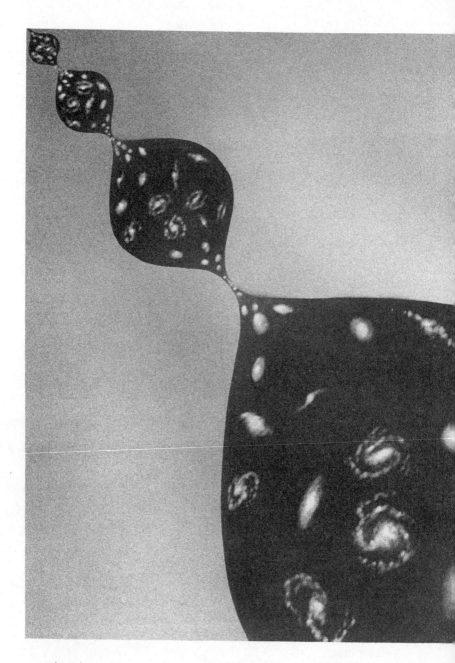

AND AGAIN

TWO

Basic Questions

ASTROPHYSICISTS, philosophers, and theologians, as well as people from all segments of society, seek to learn whether the Universe will continue to expand forever or whether its expansion will someday stop. Should there be a large enough pull of gravity exerted by all the material objects in the Universe, then the expansion might eventually halt, reversing itself into contraction. Several questions come immediately to mind: How long has the Universe been expanding? How much more time will elapse before it stops its expansion? If the Universe does start to contract, what will happen upon its eventual collapse toward a small, dense point much like that from which it began? Will the Universe just end at some finite time in the future? Or will it perhaps bounce and begin expanding anew? If the Universe has bounced before, we may well inhabit a cyclically expanding and contracting Universe—temporarily riding one of an array of cycles in a Universe having neither beginning nor end.

These are the basic large-scale fates for the Universe: it can expand forevermore; it can expand and then contract to a collapsed point and end; or it can cyclically expand and contract indefinitely. Each model represents a working hypothesis, a theory based on available data. But unless we take the final step in the scientific method and experimentally test these various models, we cannot know which one, if any, is correct.

Fortunately, we live at a time when astronomers are actually subjecting these possibilities to observational tests. Our experiments, together with the theories underlying them, seek direct answers to many of the questions above. To appreciate their findings, though, requires a deep understanding of the nature of space and time on the grandest scale. And to gain this appreciation for all space and all time, we need a particular tool. That tool is relativity theory.

Now, many people become tense and worried upon hearing the word "relativity." Relativity is surrounded by a mystique suggesting that only scientific geniuses can understand it. That might well be true at the mathematical level. But, conceptually, the foundations of relativity theory are rather straightforward— we might say relatively simple—provided we are willing to forgo terrestrial logic and human instinct.

Relativity is simple in its symmetry, its beauty, its elegant ways of describing grandiose aspects of the Universe. To be sure, it employs higher mathematics—advanced calculus and beyond— to quantify its application to the real Universe. Yet it behooves all of us to attempt to gain at least a nonmathematical feeling for some of the underlying concepts of relativity theory. In this way, we shall be better positioned to appreciate, albeit qualitatively, some of the weird physical effects we are likely to encounter while modeling the Universe, describing high-speed space flight, exploring the intricacies of black holes, and even studying the origin of all things.

Before discussing the details of relativity, however, we need to consider some of the reasons for its development, especially the mounting historical problems and inconsistencies that eventually gave rise to a whole new way of viewing our world. In this manner, we shall come to appreciate the sheer power and deep insight provided by that universal theory, often judged perhaps the greatest technical achievement of the human brain.

Sir Isaac Newton, father of the "old" gravity

THREE

Historical Problems

W E OF THE twentieth century are not the first to inquire about the nature of the Universe en masse. Certainly the Greeks of old, and perhaps civilizations before them, contemplated the grandest nature of the cosmos. Prior to the Renaissance, most earthlings were satisfied with an Earth-centered Universe. This was what Aristotle believed and what the church taught. However, with the advent of critical thinking and accurate observations in the sixteenth century, the early Greek models lost their attractiveness. The apparent motions of the Sun and planets could still be explained in terms of geometric circles —the perfect geometrical figure of Greek antiquity—but increasing numbers of circles were needed. Even today, the observed motions of all the known planets could be so modeled, though a vast number of such circles would be required to represent the orbital paths of the eight (non-Earth) planets and one Sun. Geocentric models required a complex series of large and small circles, for the planets were theorized to have circular paths

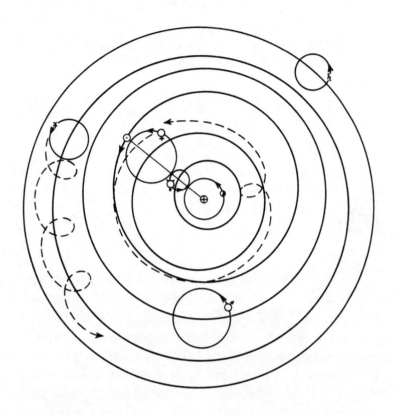

FIGURE 4 A geocentric model of the inner Solar System, based strictly on groups of small and large circles, can explain reasonably well the apparent motions of the Sun and planets around a centralized Earth. This diagram shows how, although the planetary and solar orbits are drawn to scale, they do not share a common center; the dashed lines further illustrate the complexity of such a model by depicting the resultant paths of just two planets, Venus and Jupiter, along their principal orbits. For the ancients, such a geocentric model constituted a perfectly practical system of celestial mechanics. However, modern, accurate observations of all nine planets prove that this type of model, though mathematically feasible, is incompatible with today's laws of physics. The diagrammatic symbols are as follows: ⊕ Earth, ☽ Moon, ☿ Mercury, ♀ Venus, ☉ Sun, ♂ Mars, ♃ Jupiter, ♄ Saturn.

that spun around the edges of larger circles, which in turn rolled like wheels across the sky.

Additional observations gradually made the Earth-oriented models less simple, that is, more complex. Complexity is often an adverse sign in science, for researchers trying to decipher the deepest secrets of nature almost invariably find the best models to be basically simple, beautiful, and symmetrical. When a theory or model becomes progressively more complicated, it is, more likely than not, incorrect. So people schooled in geometry abandoned the complex Greek models when, in the sixteenth century, the Polish cleric Nicolaus Copernicus proposed that a Sun-centered or heliocentric Solar System could improve the harmony and organization of the tangled geocentric models previously fashioned by the ancient Greeks. Actually, Copernicus opted for a heliocentric system because it was "pleasing to the mind," something he saw in mind's eye but could not support with any observational proof. His bold suggestion was followed by the German Johannes Kepler's declaration of the basic laws of planetary motion, the most startling of which, at least at the time, was that the planetary orbits are not circles at all but ellipses—a proposal made easier by the experimental proof then recently provided by Italy's Galileo Galilei that projectiles follow parabolic paths. Kepler's laws were empirically deduced, based solely on a lifetime's accumulation of observed data (mostly acquired by the eccentric Danish astronomer Tycho Brahe). Widespread acceptance was lacking, however, since in the sixteenth century there was no firm foundation for either the heliocentric model of Copernicus or the empirical laws of Kepler. Together, though, these new ideas did describe planetary motions accurately and simply.

A solid underpinning for the heliocentric system was secured when, in the seventeenth century, the British mathematician Isaac Newton developed a deeper understanding of the way objects move and interact. From his theory of gravitation, Newton was able to derive, using his newly invented calculus, Kepler's laws of planetary motion. These empirically deduced laws could thus be justified conceptually. Newton's monumental breakthrough gave the Copernican heliocentric model of the Solar System a great deal of mathematical support.

The principal tenet of Newtonian dynamics stipulates that, in the absence of gravity or any other force, physical objects do not change their current state of affairs: an object in a state of motion, or in a state of rest, will remain in that state unless or until it is influenced by some external force. (This statement is often termed the law of inertia, inertia being a measure of the resistance to change of motion.) For example, a pencil rolling across a tabletop or a baseball skirting along the ground should continue to roll onward forever. In reality, of course, these objects stop— that is, experience a change in motion. The friction of the tabletop or the ground eventually halts them. In this case, friction is the external influencing force. But in hypothetical free space, where there are no forces, any object will continue to move in its initially directed path forever.

The path of a planet provides a good example of Newtonian dynamics, illustrating how the planets move not in only one direction. That is, they do not fly off at tangents to their orbit, despite the fact that each planet's forward momentum would ordinarily have a tendency to cause just that. Instead, a planet's motion is "controlled" by the gravitational force of the Sun. The inward pull of the Sun, a massive object, simply prohibits each of the (less massive) planets from escaping. Accordingly, a planet's forward momentum at any point in its orbit is counteracted by the inward pull of the Sun's gravity. The resulting path is the planet's elliptical orbit around the Sun (or, more precisely, one that causes the planet and the Sun to revolve around the system's center of mass).

Planetary motions are reasonably well understood now. Astronomers have observed the planets' motions and modeled their orbits to great accuracy, and physicists have dissected the nature of the gravitational force that gives rise to those orbits. And yet,

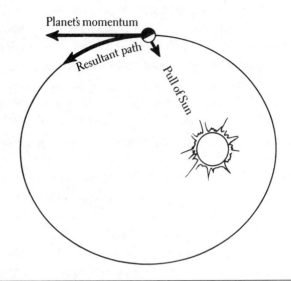

Planet's momentum

Resultant path

Pull of Sun

FIGURE 5 The motion of a planet does not follow a line tangent to its orbit, although the planet's momentum is so directed. Constant competition between a planet's forward momentum and the gravitational pull of the central Sun yields an intermediate path and thus an elliptical orbit. (A circle, by the way, is a special form of an ellipse and represents a rather improbable orbital path of one object around another.)

further thought and especially precise observations have poked some subtle holes in the Newtonian theory of gravity. Less than a century ago, models of our Solar System based on Newtonian ideas became complex—a sure sign that something was awry.

Just before the turn of the century, researchers tried to couple the force of gravity, which varies inversely as the square of the distance from an object, with the force of electromagnetism, which also varies inversely as the square of the distance. The identical mathematical forms of the essential laws of gravity and of electromagnetism suggest a common relationship; perhaps they are different manifestations of the same basic force (much as the Scottish physicist James Clerk Maxwell had earlier shown regarding electricity and magnetism themselves). However, when attempts were made to merge them, other problems quickly arose.

First, the forces of gravity and electromagnetism seemed (at least at the time) to transfer information differently. To conceptualize this, imagine two electrically charged particles, one positive and the other negative, as well as the electromagnetic forces extending outward from each particle. Should one of these charges wiggle or move in any way, the other particle would detect that motion. But each charge would respond to the other's only after a certain amount of time. This is true because electromagnetic information moves at the fast, but finite, velocity of light. Electromagnetism does not act instantaneously. By contrast, the action of gravity according to Newton is virtually instantaneous, having an effect that is immediate and without delay.

Another basic difference between these two similarly constructed forces concerns gravity's literal reach outward from *every* massive object to the limits of the observable Universe. The gravitational pull of even small bodies—like people, or even atoms—never diminishes to zero. In other words, every object having any mass exerts some net gravitational force on every part of the Universe. The effect of gravity simply cannot be canceled. Not even large clumps of antimatter (should they exist in the Universe) can counter gravity. Antimatter does not cause antigravity or any other kind of magical, repulsive force. Gravitational influence is always one of attraction, regardless of whether the object in question is made of matter, antimatter, or a mixture of the two.

Electromagnetism, by contrast, *can* be canceled. A group of equal numbers of positive and negative charges results in an electromagnetic force of zero. Furthermore, that force is sometimes attractive (for oppositely charged objects) and at other times repulsive (for like charges). The important point is that the net effect of electromagnetism can be altered, depending upon the mixture and positioning of positive and negative charges. This is not true of gravity.

Other fundamental differences span the theories of gravity and electromagnetism, among them the nature of the forces caused by moving objects. In basic physics, we learn that an electrically charged particle in motion induces a magnetic force. What's more, physicists have known for a century or so that the opposite is also true: a magnet in motion induces an electric force (this is the principle of electric dynamos). However, Newton's gravitational force has no dependence on motion. A massive, moving object neither induces another type of gravitational force nor alters its own.

When researchers tried to unify the theories of gravity and electromagnetism—simply on the basis of their identical adherence to the inverse-square law—they quickly encountered these and other basic differences. The two forces did not then appear to have much in common after all (at least not in our terrestrially familiar environment). These studies prompted physicists to attach even greater significance to that special velocity with which electromagnetic information is exchanged—the velocity of light.

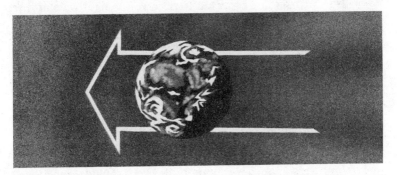

FIGURE 6 Imaginative sketch of Earth's motion through the "universal ether." This ether was once thought to be a material medium through which light waves travel like vibrations in a bowl of jelly, in analogy to the air through which sound waves are transmitted or to the water through which ocean waves propagate. Now known to be nonexistent, the ether is also unnecessary, since modern physics demonstrates that electromagnetic radiation can travel through a vacuum.

By using the velocity of light as a standard, scientists of nearly a century ago reasoned that it might be possible to measure the net motion of Earth through the Universe. In other words, the velocity of light itself might be a benchmark against which Earth's *absolute* motion could be measured relative to the rest of the Universe. In particular, Earth's motion could then be documented through what nineteenth-century astronomers imagined to be a "universal ether"—a stationary, substantive, but invisible medium that permeated all of space. (We now recognize that no such ether exists; not even interstellar gas is motionless or all-pervading.)

About a century ago, the American physicist Albert Michelson and his chemist assistant, Edward Morley, performed a key experiment to test the ether's suitability as an absolute frame of reference so crucial to Newton's mechanical worldview. This experiment, as simplified in figure 7, required two beams of light radiation to be launched in opposite directions. In one direction, Michelson and Morley expected to observe the light traveling at a rate equal to the velocity of Earth plus the standard velocity of light. In the opposite direction, they expected to observe the difference between the velocities of light and of Earth. These are the straightforward predictions of Newtonian dynamics and of common sense.

The situation here is akin to that of two northbound automobiles traveling with velocities of thirty and forty kilometers/hour. Should they collide, little damage is done, their relative velocity being only ten kilometers/hour. However, should there be a collision between northbound and southbound cars having those velocities, the damage is considerable, their relative velocity then being seventy kilometers/hour. More to the point, a bullet fired straight ahead from a gun mounted atop a speeding aircraft has a much greater velocity than one shot from the same gun fixed on the ground; in fact, the ground speed of the aircraft's bullet equals the sum of the normal velocity of the bullet plus that of the moving plane. If this type of reasoning is valid, then we should be able to measure Earth's absolute motion relative to the ether by comparing our planet's motion to that of a beam of light.

However, the result of the Michelson-Morley experiment was a complete surprise. It not only defied common sense but also led to the downfall of the Newtonian ideas of motion and of gravity. Here's the surprise: regardless of how or from what direction the light beams were viewed, the beams traveled only at the velocity of light, neither more nor less. Observers to the left of figure 7 did not measure more than the standard velocity of light; nor did observers to the right in figure 7 measure less than the standard

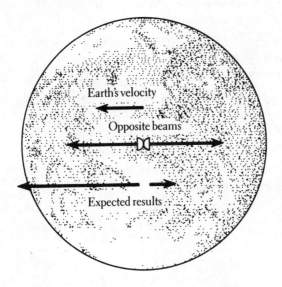

FIGURE 7 Schematic diagram of the *essence* of the Michelson-Morley experiment. If planet Earth is moving to the left, as depicted in this figure, then external observers on the left might reasonably be expected to measure a faster-moving beam of light than observers on the right. But the experiment proves categorically that they do not, that Newtonian ideas and human intuition go awry when the analysis involves the velocity of light.

velocity. Instead, all observers measured the velocity of light—and *only* the velocity of light. Apparently, the velocity of a light beam is unaffected by the speed of the object that releases the beam.

The key result of the Michelson-Morley experiment has been confirmed many times during the past century, an elegant and recent version being that light emitted from both limbs of the rotating Sun travels to Earth with the same velocity. (Sunlight is red and blue shifted, in accord with the Doppler effect as discussed in chapter 1, but the speed of the light is unaffected by the speed of its source.) This experiment, in all its forms and modern versions, is a classic test of contemporary physics, and it has forced researchers to conclude that the velocity of light is more than important and special, that it must be fundamental. It is an absolute physical constant, perhaps the most important quantity in all of physics.

Aside from a host of untenable technical proposals made throughout the twentieth century, there are two principal solutions to this observational dilemma. Either Earth has no motion at all through the Universe, which is nonsense, since measurements of the annual back-and-forth wobble of nearby stars (stellar parallax) prove that Earth moves around the Sun. Or the fundamental basis of Newtonian dynamics—especially the common-sense computation of relative velocity—must be incorrect, thus making it impossible to measure Earth's absolute motion through space.

RELATIVITY THEORY

The successes of the Maxwell-Lorentz theory have given great confidence in the validity of the electromagnetic equations for empty space and hence, in particular, to the statement that light travels "in space" with a certain constant speed c. Is this law of the invariability of light velocity in relation to any desired inertial system valid? If it were not, then one specific inertial system or more accurately, one specific state of motion (of a body of reference), would be distinguished from all others. In opposition to this idea, however, stand all the mechanical and electromagnetic-optical facts of our experience.

For these reasons it was necessary to raise to the degree of a principle, the validity of the law of constancy of light velocity for all inertial systems.

—A. Einstein, *Out of My Later Years*

Einstein at his desk in the Patent Office, Bern, 1905

FOUR

The Special Theory of Relativity

SCIENTISTS at the turn of the century were thoroughly puzzled by the enigma presented by the Michelson-Morley experiment, until a German-Swiss patent examiner named Albert Einstein offered a startling explanation. In doing so, he first rejected the idea of an ether as a fixed reference frame and then conceived a mathematical theory that resolved the conceptual problems of measuring precisely the velocity of light—never more, never less —regardless of when, how, and from what angle the light is observed. Einstein's hypothesis is called relativity theory, or just "relativity," for short.

Relativity theory has two main tenets, both enunciated by Einstein in 1905; one does away with the concept of absolute space, the other with that of absolute time. Together, they lead to history's most famous equation, $E = mc^2$, where E, m, and c represent energy, mass, and the velocity of light, respectively.

The first tenet of relativity is that the basic laws of physics are the same everywhere and for all observers, regardless of where the persons are or how fast they might be moving. (The gist of this principle, as it applies to mechanical laws, actually dates back to the Renaissance, when Galileo first proposed it and Newton himself later championed it.)

The second tenet of relativity (and one utterly foreign to a Galileo or a Newton) is that there exists a fourth dimension— time—which is equivalent to the usual three spatial dimensions. An object's position can be generally described by three dimensions of space—either right or left, either up or down, and either in or out; three dimensions are sufficient to describe *where* any object is located in *space*. A fourth dimension of *time* is needed to describe *when*— either past or future—an object exists in that space. By coupling time with the three dimensions of space, Einstein was able to reconcile long-standing, though subtle, inconsistencies regarding Newton's ideas about nature; above all, he was able to explain the results of the Michelson-Morley experiment by postulating that the numerical value of light velocity is an absolutely constant number at all times and to all observers, regardless of when, where, or how light is measured. In fact, issues of space and time are so thoroughly intertwined in Einstein's view of the Universe that he urged us to regard these two quantities not as space *and* time but as one—*spacetime*.

The concept of spacetime is not terribly difficult to understand. We regularly encounter spacetime notions throughout our lives, often without recognizing them. After all, when arranging

---

FIGURE 8 A sample page, written in Einstein's hand, of perhaps the most revolutionary scientific paper ever published. "On the Electrodynamics of Moving Bodies," by A. Einstein, appeared in the 1905 edition of *Annalen der Physik*. It proposed, among other things, the abolition of absolute space and time, the equivalence of mass and energy, and all manner of baffling alterations regarding the relative measurements of length, mass, and time.

bei α eine vorläufig unbekannte Funktion $\varphi(v)$ ist und der Kürze halber angenommen ist, dass im Anfangspunkt von k für $\tau = 0$ $t = 0$ sei.

Mit Hilfe dieser Resultate ist es leicht, die Grössen ξ, η, ζ zu ermitteln, indem man durch Gleichungen ausdrückt, dass sich das Licht (wie das Prinzip der Konstanz der Lichtgeschwindigkeit in Verbindung mit dem Relativitätsprinzip verlangt) auch im bewegten System gemessen mit der Geschwindigkeit V fortpflanzt. Für einen zur Zeit $\tau = 0$ in Richtung der wachsenden ξ ausgesandten Lichtstrahl gilt

$$\xi = V\tau$$

oder

$$\xi = \alpha V\left(t - \frac{v}{V^2 - v^2} x'\right)$$

Nun bewegt sich aber der Lichtstrahl relativ zum Anfangspunkt von k im ruhenden System gemessen mit der Geschwindigkeit $V - v$, so dass gilt

$$\frac{x'}{V - v} = t$$

Setzen wir diesen Wert von t in die Gleichung für ξ ein, so erhalten wir:

$$\xi = \alpha \frac{V^2}{V^2 - v^2} x'.$$

Auf analoge Weise erhalten wir durch Betrachtung von längs den beiden andern Axen bewegten Lichtstrahlen:

$$\eta = V\tau = \alpha V\left(t - \frac{v}{V^2 - v^2} x'\right),$$

wobei

$$\frac{y}{\sqrt{V^2 - v^2}} = t \ , \ x' = 0 \ ;$$

also

$$\eta = \alpha \frac{V}{\sqrt{V^2 - v^2}} y$$

und

$$\zeta = \alpha \frac{V}{\sqrt{V^2 - v^2}} z \ .$$

Setzen wir für x' seinen Wert ein, so erhalten wir

$$\tau = \varphi(v)\,\beta\left(t - \frac{v}{V^2} x\right)$$

$$\xi = \varphi(v)\,\beta\,(x - vt)$$

$$\eta = \varphi(v)\, y$$

$$\zeta = \varphi(v)\, z$$

wobei $\beta = \dfrac{1}{\sqrt{1 - \dfrac{v^2}{V^2}}}$

to meet someone, we must specify not only the place but also the time. Otherwise, we would never rendezvous successfully in order to create the event of meeting at the same place *and* at the same time.

Hermann Minkowski, a German mathematician and one of the early champions of Einstein's novel ideas (which were much resented in the physics community initially), put it succinctly in a memorable lecture in 1908: "From this hour on, space as such and time as such shall recede to the shadows and only a kind of union of the two retain significance."

To appreciate how the concept of time can be coupled with space, consider the following example. Figure 9 is a diagram (restricted to the two-dimensional plane of the page) in which time is plotted along one axis, while the three dimensions of space are consolidated along the other axis. A plot of this sort is called a spacetime reference frame. Such diagrams provide a compact way of describing the position or motion of an object in the realm of four-dimensional spacetime. Plots like these must be used because there is no clear way to draw in four dimensions. Of course, mathematics can be used to describe four (or any number of) dimensions, but no person—not even the smartest mathematician—can visualize four dimensions. Even three dimensions cannot be easily illustrated on a two-dimensional flat surface like the page of a book. So, for convenience, we compact the three dimensions of space onto one (horizontal) axis. In that way, time can become the other dimension on the second (vertical) axis.

The spacetime path drawn in figure 9 depicts a typical sequence of events that occurred in space and time. First, it shows at the upper right an event that describes, for instance, the start of a lecture. That event, labeled no. 4 in the figure, occurred in some space, say, a classroom, and at some time, say, 1300 hours on a particular day. To get there—that is, to experience this event of the start of the lecture—we can imagine that students, a kilometer away, might have initially departed for class at, say, 1245. This event, the act of departing for class, is described by the point marked no. 1, at the lower-left corner of the spacetime diagram. For the students to rendezvous successfully with the lecture event, which was still at that time in their future, they had to travel the kilometer-long route of space in fifteen minutes of time.

For example, the student whose spacetime diagram is shown in figure 9 first walked at a steady pace after leaving her apartment at 1245. In doing so, she was able to cover a substantial amount of space in a certain amount of time. The figure specifies that she traversed one-third of a kilometer in six minutes, after which she

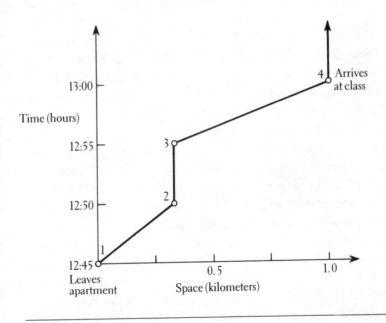

FIGURE 9 A spacetime diagram is used to denote a series of events that occur somewhere in space and somewhen in time. This one is meant to show how a student, having left her residence at 1245, sporadically makes her way to class some fifteen minutes later. The labeled events are (1) leaving for class, (2) stopping to talk to a friend, (3) resuming the trek, and (4) arriving at class.

noticed a friend. Stopping and talking takes some time too (five minutes shown here), and this is also an event, labeled no. 2 in the figure. During this five-minute interval, her velocity was zero (as represented by the vertical line in figure 9), and she traversed no space. It is now 1256, four minutes prior to the lecture. Lest she be late for class, the student must cover the remaining amount of space in a rather small amount of time. In other words, her velocity must increase (alas, she will need to run). As indicated by the final segment of the line drawn in figure 9 (event no. 3 to event no. 4), this increase in velocity requires the student's path through spacetime to become more horizontal.

The above description is a simplified example of how physicists analyze various events occurring all around us in space and time. Often they draw more general spacetime reference frames capable of describing velocities much larger than those attainable by students walking to class. Figure 10 shows such a spacetime reference frame, once again diagramming the time axis vertically and the space axis horizontally. Here, the scale differs greatly from that of figure 9; an interval of 300,000 kilometers of space equals an interval of one second of time. Accordingly, the velocity of light—300,000 kilometers/second—is represented by a 45° line, shown dashed in figure 10. The origin of the diagram, where the axes intercept, represents "here" in space and "now" in time.

All the events with which we are familiar in our practical experience cover a relatively small space in a given amount of time. That's because our everyday velocities are small—very small —compared with that of light. Consequently, familiar events involving humans are represented by lines very close to the vertical time axis of figure 10. In other words, events that happen slowly are plotted near the time axis, whereas very fast events are plotted near the 45° dashed line.

Other aspects of spacetime diagrams are interesting as well. For example, since no event can occur more quickly than the velocity of light, the only paths permitted on such a spacetime

diagram are those within the unshaded cone defined by the 45°
dashed lines. Within the top cone lies the future; within the
bottom cone, the past. Any event destined to occur tomorrow,
next week, or next year—that is, at any time in the future—must
necessarily be plotted inside the upper portion of the unshaded

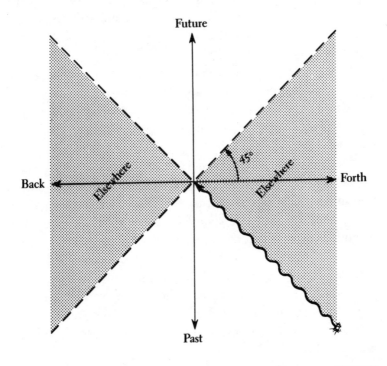

FIGURE 10 A more general spacetime diagram, in which any event can be
plotted. The center of the diagram marks "here" and "now." All other events
denote "there" and "then." The dashed diagonal lines define a "light cone"
inside of which realistic events can be plotted; the bottom of the cone represents
the past and the top the future. The shaded area outside the cone is termed
"elsewhere," since events so plotted in that domain of the diagram would
necessarily occur at velocities greater than that of light, a proposition considered
improbable given the currently known laws of physics. The wavy line *(lower
right)* denotes a typical spacetime path of a star's light moving from the far away
and long ago to the spacetime "present," where it can be detected.

cone. Conversely, information from past events, like the emission of light from cosmic objects millions of light-years away, must travel from the past toward the origin along one of the 45° lines describing the cone. While observing an astronomical object, we see its radiation arriving here at our location and now in our time. The event describing that object's emission can be plotted if the object's distance is known. The greater the distance from us, the farther that event—the act of emitting radiation—is plotted from the origin in such a spacetime diagram.

This is why astronomers contend that looking out into space is equivalent to probing back into time. The two—space and time—are intimately coupled. Looking out is looking back. Telescopes are time machines; astronomers are historians. Together we probe the past.

Two notes of caution are in order. The first is that no event can occur within the shaded area of figure 10. Any event outside the unshaded cones enclosed by the 45° dashed lines would require more than 300,000 kilometers to be traversed in a single second. That's equivalent to traveling at a velocity greater than that of light, and no one has ever detected material quantities traveling at such truly high velocities. Einstein's theory demands that no object accelerate beyond the velocity of light. Accordingly, the shaded area beyond the unshaded cone is often labeled "elsewhere"; it is the regime of forbidden events—a twilight zone of sorts.

The second caveat is that all new events occur in the forward time domain, in the direction of the future. Interestingly enough, we shall see in the next chapter that relativity permits movement into the future at rates higher than those allowed by common sense. But events cannot, under any circumstances, transpire toward the past. Travel into the past would violate a basic principle of cause and effect: the cause of an event must precede the effect of that event. Hence, we cannot travel backward in time; otherwise, we could potentially experience the event of our existence prior to, for example, the birth of our parents. Voyages into the past violate not only the philosophy of cause and effect but one of the central dogmas of biology as well.

So all those science-fiction stories that routinely envision travel into the past are sheer flights of fancy. We can *observe* the past by "looking out," but the laws of physics absolutely prohibit journeys into the past. Travel into the future, however, even at unbelievably high speeds, is not prohibited by any such cause-and-effect arguments.

Incidentally, most physicists maintain that no object of any type can travel faster than the velocity of light. Contrary to popular belief, however, this statement is not a direct conclusion of relativity theory. In fact, some theorists have stressed that Einstein's ideas in no way prohibit the existence of superluminal (faster-than-light) objects. To account for such superswift objects, a few unorthodox researchers have recently proposed three general classes of objects.

The first class comprises slow-moving objects of the everyday world around us—atoms, molecules, planets, stars, people, and so on. All scientists agree that this class exists. And all agree that relativity predicts weird observational consequences whenever any of these objects travel close to the velocity of light (as we shall see in the next chapter). Furthermore, all researchers agree that these often sluggish objects cannot be accelerated beyond the light barrier.

A second class of objects, exclusively subatomic particles, move only *at* the velocity of light. For example, scientists have never detected photons or neutrinos traveling at more or less than the velocity of light. These particles are not accelerated up to that special velocity; instead, they are instantaneously created in atomic and nuclear reactions, after which they immediately zip away at the velocity of light. Nor do they slow down or speed up.

A question is often raised: Can relativity theory accommodate a third class of objects—superluminal objects that move with speeds *only larger* than that of light? The mathematical answer seems to be that there *could* indeed exist an entire family of particles that travel faster than light, provided they never attempted to slow down too much. In other words, these particles would also find the velocity of light to be a limiting value or barrier, *slower* than which they could not travel. Such particles are called tachyons, after the Greek word ταχισ, meaning "swift."

Currently, we have no evidence whatsoever for the existence of tachyons, despite much experimental effort to detect them. Theoretically, though, nothing in relativity theory seems to exclude them. What relativity does clearly prohibit is crossings of the light barrier. Accordingly, no object in our "slow-moving," everyday world could ever travel faster than light, whereas no tachyon could ever travel slower! If tachyons do exist, they will be fascinating to study, for their spacetime properties will lie in that unworldly portion of figure 10 mysteriously labeled "elsewhere."

Return for a moment to the Michelson-Morley experiment. To see how relativity theory can explain the peculiar results of this basic experiment, consider two spacetime reference frames. Let one frame have velocity 1 and the other velocity 2. These velocities can be of any value provided they do not exceed that of light. The two frames can represent any objects—for instance, elementary particles, people, rocket ships, or galaxies. For convenience, imagine these two frames to be approaching one another.

Now ask, What is the relative velocity of the two reference frames? Common sense (and Newtonian theory) predicts that the relative velocity is the sum of the two individual velocities. For example, if our frames represent two cars, one traveling at forty kilometers/hour and the other at sixty kilometers/hour, then the relative velocity is the sum of the two velocities, or one hundred kilometers/hour. The relative velocity of the two frames can be quantified rather simply:

$$\text{relative velocity} = \text{velocity } 1 + \text{velocity } 2$$

This is nothing more than common sense gained from everyday experience with moving objects. Descriptions of all motions, analyzed by the laws of Newton, derive from hardly more than human intuition, for these are the concepts inherent in today's culture. Reasoning of this type is completely valid in our everyday world, that is, in realms of velocities much smaller than the velocity of light.

FIGURE 11 Two spacetime reference frames approaching each other, one with velocity 1, the other with velocity 2.

The above equation for the relative velocity cannot be completely correct, however. Why not? Well, suppose two frames of reference move toward one another with much, much higher velocities. If the above equation is correct, the relative velocity can then exceed that of light. For example, suppose the two reference frames of figure 11 are rocket ships traveling at velocities of 150,000 kilometers/second and 200,000 kilometers/second. Each ship travels below the velocity of light, but the sum of their velocities—the relative velocity in Newtonian theory—would then be 350,000 kilometers/second. And this is a good deal faster than the 300,000 kilometers/second with which light or any other type of radiation travels.

Another, even more extreme case arises when each of the two reference frames moves with a velocity *equal* to that of light itself. This could be the case for light photons themselves, microscopic bundles of energy traveling head-on toward one another. If the above equation is correct, we will expect their relative velocity to equal twice the velocity of light. But this analysis must be incorrect, for nothing (save the strictly hypothetical tachyon) can exceed light velocity. The Michelson-Morley experiment and numerous experiments since have clearly demonstrated that no velocity—not even a relative velocity—can exceed that of light.

Something, therefore, is wrong with the above commonsense equation. While it seems to work properly for small velocities, it fails for very high velocities. Fortunately, Einstein was able to unravel this dilemma. He showed that the correct relative velocity of two approaching reference frames is really a slightly altered version of the above equation. The alteration takes the form of a factor by which the previous equation must be divided. In other words, the correct relationship becomes

$$\text{relative velocity} = \frac{\text{velocity 1} + \text{velocity 2}}{1 + \dfrac{\text{velocity 1} \times \text{velocity 2}}{(\text{velocity of light})^2}}$$

The denominator on the bottom of this revised equation derives from a detailed mathematical analysis of the nature of space and time. An aspect of this factor appears throughout the theory of relativity, affecting almost all physical quantities therein. Some researchers call it the noncommonsense factor; others label it the relativistic factor. Whatever the name, it is the factor that alters our commonsense notions of natural phenomena. And it is responsible for some rather bizarre results that clearly violate human intuition, especially when the velocities involved are comparable to that of light.

The mathematical form of this relativistic factor should not frighten anyone. Note that the factor becomes important only when objects move extremely fast, as can be seen by closely examining it;

$$1 + \frac{\text{velocity 1} \times \text{velocity 2}}{(\text{velocity of light})^2}$$

Note that this factor nearly equals 1 whenever the velocities 1 and 2 are small. The square of the velocity of light is a truly huge number, and, under these circumstances, the right-hand term of the above factor is virtually zero. For all practical purposes—namely, in the realm of ordinary, everyday things—Newtonian physics is completely valid. When objects travel much below the velocity of light, the Einsteinian equation for the relative velocity reverts to the Newtonian form of (velocity 1 + velocity 2). We can prove this by setting the relativistic factor equal to 1 in the Einsteinian equation for relative velocity. The upshot here is that the commonsense notion of velocity addition is perfectly fine (that is, an excellent approximation of reality), provided objects travel much more slowly than the velocity of light.

When an object's velocity is very large, however, the relativistic factor does not equal 1; it becomes larger. How much larger depends on the actual velocity. Departure from commonsense notions begins to occur when the velocity approaches that of light. In fact, only when an object's velocity is within some 10 percent of the velocity of light does the relativistic factor become significantly greater than unity.

The beauty of the relativistic factor lies in its ability to explain the central, yet at face value puzzling, result of the Michelson-

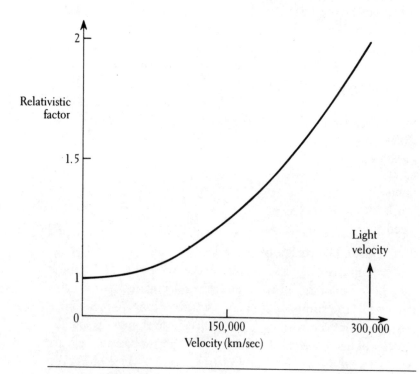

FIGURE 12 A plot of the relativistic factor against velocity. Note how this factor equals 1 (or nearly 1) for small and moderate velocities but grows larger when the velocity approaches that of light.

Morley experiment. It also enables us to justify the extreme hypothetical case of two spacecraft that approach one another with velocities equal to that of light. With the correct, Einsteinian equation for the relative velocity, the result is not twice the velocity of light, as expected from Newtonian physics. Rather, it is

$$\text{relative velocity} = \frac{\text{velocity of light} + \text{velocity of light}}{1 + \dfrac{\text{velocity of light} \times \text{velocity of light}}{(\text{velocity of light})^2}}$$

or,

$$= \frac{\text{twice the velocity of light}}{1 + 1}$$

or,

$$= \text{velocity of light}$$

In other words, the maximum-possible relative velocity equals the velocity of light. Likewise, the Einsteinian equation yields a relative velocity of only about 260,000 kilometers/second for the previously discussed rockets traveling with velocities of 150,000 and 200,000 kilometers/second. Hence, in Einsteinian physics, but not in Newtonian physics, the relative velocity is always equal to or less than that of light. Nothing has ever been observed to exceed this absolute speed limit.

CURIOUSER AND CURIOUSER

Strange Consequences

AS WAS MENTIONED earlier, the relativistic factor inundates the theory of relativity. It, or some form of it, is attached to virtually every physical quantity describing an object or an event within our four-dimensional Universe. Accordingly, an object behaves as expected on the basis of common sense only if its velocity is small compared with that of light. But once the relativistic factor begins to depart from 1 as the velocity of any object approaches that of light, some demonstrably weird effects can occur. In this chapter, we consider some of the baffling consequences of relativity theory.

Imagine a spaceship traveling past us with a velocity typical of those man-made spacecraft now exploring our Solar System — namely, much more slowly than the velocity of light. In terms of the concepts discussed earlier, this case stipulates a moving spacetime reference frame (the spaceship) passing a stationary reference frame (us). (By the way, all material objects engineered thus far by humans travel much, much below the velocity of light. The fastest-moving craft to date is the U.S. *Pioneer 11* robot, which recently escaped the Solar System with a velocity of some 60 kilometers/second, or 0.02 percent of the velocity of light; satellites orbit Earth with velocities of about 8 kilometers/second, and

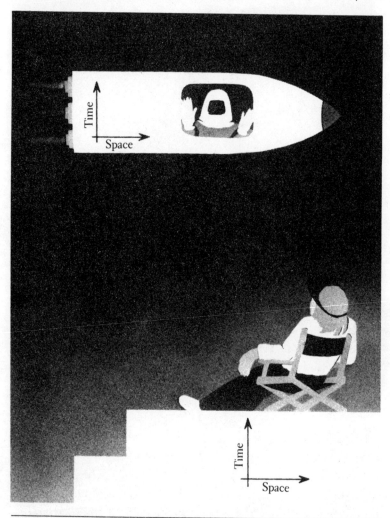

FIGURE 13 The case of a moving spacetime reference frame passing a stationary reference frame.

high-speed cannon projectiles achieve merely 2 kilometers/second. Even Earth's motion around the Sun measures "only" 30 kilometers/second and the Sun's around the center of the Milky Way Galaxy some 250 kilometers/second, the latter being hardly a tenth of one percent of the velocity of light.)

In principle, we could devise an experiment to measure several physical quantities describing the spaceship. For instance, the size of the ship could be determined, as could its mass. Provided the experiment is done carefully, our estimates of the ship's physical quantities (while the ship passes us rather slowly) would match the values measured by the travelers on board the spaceship. Everyone would agree, and common sense would prevail, because the relativistic factor equals 1, for all practical purposes.

On the other hand, if the spaceship were to pass us with a much higher velocity, say, with a motion close to the velocity of light, the relativistic factor decidedly affects the outcome. As a result, our estimate of the spaceship's length would no longer agree with those measured either by the on-board travelers or by us when the spaceship passed more slowly. Instead, we stationary observers would measure a smaller spaceship; its length would have shrunk. The greater the spaceship's velocity, the greater the amount of shrinkage. This effect is not an optical illusion. Nor should we try to find a mechanical cause for the shortening. It is the spaceship's motion alone that causes the ship to be foreshortened in the direction of motion—not an overall shrinkage of the entire ship's size, just in its direction of motion.

Figure 14 plots the extent of shrinkage as a function of relative velocity. The curve shows, for example, how a typical meter stick progressively shrinks in length, from the usual meter when the stick travels at low velocities to much smaller values while traveling very rapidly. Should the stick attain the velocity of light itself, it would diminish to nothing at all, once again implying the limiting nature of the velocity of light. Of course, it is currently impossible for our rather primitive technology to launch space-

ships and meter sticks to anywhere near that of the velocity of
light. The curve of figure 14 is merely meant to predict how we,
as stationary observers, would experimentally perceive a rapidly
moving object. Recognize, though, that this contraction in length
is strictly a consequence of the relativistic factor applied to the
measurement of size and shape. It is totally independent of the
composition of the meter stick, for it could be made of wood,
metal, plastic, or whatever.

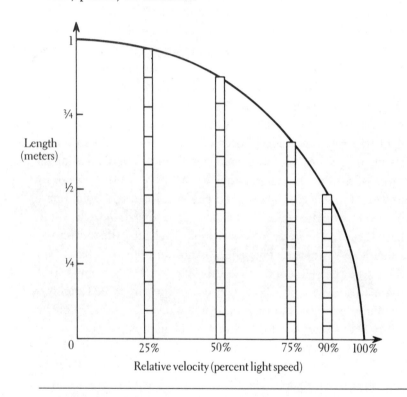

FIGURE 14 This curve shows how a stationary observer would note a shrinkage
in the length of a moving meter stick; at small relative velocities, the meter stick
equals a full meter, but at large velocities it is foreshortened. Specifically, such
a stick moving at 90 percent of the velocity of light would shrink to about half
a meter. (The horizontal scale is expressed as a percentage of the velocity of
light.)

The most intriguing aspect of this weird effect is that the effect itself is relative. In other words, the travelers on board the spaceship experience no change either in the length of their spaceship or in any meter stick that may be on board. This is true regardless of the ship's velocity. Why so? Because the travelers are in their *own reference frame:* the on-board travelers have zero velocity relative to their spaceship reference frame. They measure an on-board meter stick to be precisely one meter long because they are traveling with it. The travelers share the same reference frame as the on-board meter stick, as well as the reference frame of the spaceship itself. There is no relative motion between the on-board travelers and their spaceship, or between the on-board travelers and their on-board meter stick. Thus, the travelers deduce a relativistic factor of unity, provided they examine objects within the spaceship. But once they look out of the ship's window, their perspective changes. Now they see their motion relative to the outside world and hence deduce a different relativistic factor. In other words, as the on-board travelers peer through the spaceship's windows at the rest of us standing stationary (see figure 13), they measure all objects to be smaller in the direction of motion. *They* see *us* as thinner individuals. And, if we are holding a meter stick horizontally, they will measure it to be less than a meter.

You may want to ask, Who is correct? Which objects are really shortened—the ones in the reference frame on the spaceship or the ones in our own? The answer is that both observers are correct. The bizarre consequences of relativity theory apply to any and all observers, provided they share some large relative velocity. Measurements of length depend upon the reference frame from which they are made. It's all a matter of relative velocity—hence the name "relativity theory."

Remember, though, these special relativistic effects occur only when relative motions approach the velocity of light. As can be noted from figure 14, even velocities equal to half that of light

—namely, 150,000 kilometers/second—are hardly sufficient to produce noticeable foreshortening in the direction of motion.

Length is not the only physical quantity changed by very high relative velocities. The relativistic factor also has an effect on mass. If the spaceship sketched in figure 13 were moving rapidly —say, close to the velocity of light—we, as stationary observers, would measure its mass to be greater than when it was at rest. In principle, we could set up an experiment and physically measure the ship's mass to have increased. The greater the ship's velocity relative to us, the larger the additional mass measured by us. Even the travelers on board become more massive from the viewpoint of a stationary observer.

Like length, the physical mass is unchanged from the perspective of the travelers on board the spaceship. They do not measure their rapidly moving spaceship to be more massive than usual, and they themselves appear normal. That's because, relative to their spaceship, their velocity is zero and their relativistic factor is 1. They are correct in viewing their spaceship as normal, but we are also correct in viewing it as grossly overweight. Measurements of mass, like those of length, depend on the relative velocity of the reference frame from which they are measured.

Where does the increased mass come from when an object moves at high speed? From energy. It is on this basis that Einstein arrived at a conclusion of immense importance to the world. He reasoned that since the mass of a moving body increases as its motion increases, and since motion is a form of energy (that is, kinetic energy), the increased mass of a moving body must result from its increased energy. In short, energy must have mass, and in a few mathematical manipulations he proved it: $E = mc^2$. Matter and energy are interchangeable. More than that, they are equivalent, as was demonstrated in 1945 at Alamogordo, New Mexico, when a small piece of uranium matter was transmuted into a vast quantity of light, heat, sound, and motion—which we call energy.

Incidentally, these changes in length and mass serve to limit severely the prospects for interstellar spaceflight. Since powered spaceships of the future will not be able to carry enough fuel for long-duration flights, they will need to scoop fuel from interstellar

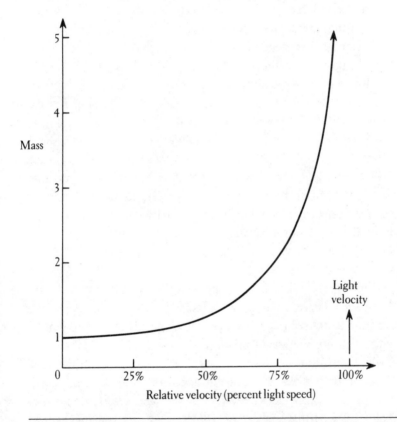

FIGURE 15 Mass of a moving object, as measured by a stationary observer, increases with the relative velocity of that object. Like its length, the object's rest mass shows clear deviations at "merely" a few tenths of the velocity of light, and its mass is perceived to change quite drastically when the relative motion approaches the velocity of light.

space while cruising along. But, from the viewpoint of an outside reference frame—namely, that of interstellar space—a rapidly moving spaceship will gradually grow more massive *relative to interstellar space*. The faster the ship tries to go, the more fuel it needs—not just because the ship is accelerating, but also because such a growing ship has a larger appetite. Even more exotic techniques not reliant upon powered rocketry, such as starship sailing using the radiation pressure of starlight or beamed propulsion using energy generated in our Solar System and directed toward a passive starship, are adversely affected; as the ship speeds forward, it grows more massive relative to its distant power source, thus requiring ever-greater energies the faster the ship moves. High velocities are therefore self-defeating. Accordingly, the relativistic effect of increasing mass at high velocities is sure to restrict interstellar spaceflight even for the most technologically advanced civilizations. Contrary to popular belief, relativity theory predicts as virtually impossible travel over truly galactic distances within a single human lifetime. However, relativity does not in any way prohibit long-duration voyages, provided the journey is a slower one, spanning many generations of human life.

Finally, the relativistic factor introduced by Einstein to account for relative motions also affects time. For example, if we, as stationary observers, could see a clock on board a rapidly moving spaceship, we would measure the clock to be ticking more slowly than normal. This effect has nothing to do with the construction of the clock, because it applies equally to pendulums, hourglasses, spring clocks, digital watches, and even (presumably) biological rhythms such as heartbeats. Nor are the mechanical workings of the clock affected by motion; only our perception of the fast-moving clock is affected. In the hypothetical case of a spaceship traveling *at* the velocity of light, we would not be able to detect any ticking at all of an on-board clock. Relative to us

—that is, from our viewpoint—time on the spaceship would have stopped altogether.

Figure 16 shows how a one-second interval of spaceship time grows larger from our viewpoint as stationary observers. If the ship were moving at the ultimate velocity, that of light itself, we would theoretically perceive a one-second interval to be infinitely long.

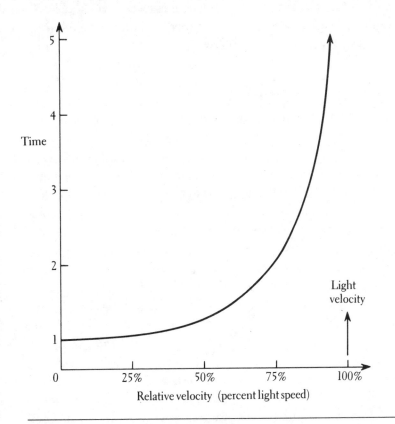

FIGURE 16 The duration of a second-long tick of a spaceship's clock increases as the relative velocity between the ship and an external observer grows large. If the relative velocity (hypothetically) reached the speed of light, time would halt altogether and, at least from the perspective of the external observer, the ship's crew would achieve immortality!

Equivalently, we could say that, from our viewpoint, our colleagues aboard the craft had effectively achieved immortality. Not surprisingly, however, the on-board travelers experience no strange timing problems; their clocks tick normally because the clocks on board have zero velocity relative to their spaceship. Only when the travelers look at us through their ship's windows do *they* measure *our* clocks to be operating more slowly than normal. Like those of length and mass, measurements of time depend on the relative velocity between observers.

A famous case serves to illustrate the phenomenon of time extension. Harry and Tom are twins, twenty years old and residents of planet Earth at the start of the experiment. Now, suppose Harry is launched toward a star system some twelve light-years away. If the voyage could be made at precisely the velocity of light, neglecting such practical matters as launch, turnaround, and landing back on Earth, the round-trip would take twenty-four years. But only in our wildest dreams can we imagine that technology would ever actually permit space travel *at* the velocity of light. To be sure, no massive object can attain the velocity of light and remain intact; all of its mass would become pure energy via $E = mc^2$, and the material object would cease to exist.

Consider, then, a slower trip. Allow Harry to travel toward the star at a velocity, say, 60 percent of the velocity of light. Since this equals $\frac{3}{5}$ of the velocity of light, the round-trip would then take twenty-four years times $\frac{5}{3}$, or forty years. That is, from the viewpoint of Tom, who remained on Earth, the complete voyage would have taken forty years. So when Harry returns, Tom would be twenty years plus forty years, or sixty years, old. But upon return Harry would not be sixty. He would be younger.

From the viewpoint of the stationary observer, Tom, the clocks on board Harry's speeding spaceship would tick more slowly than normal—that is, more slowly than clocks Tom had on Earth. Tom perceives Harry to grow old at a slower-than-normal pace, so that upon return Harry is younger than Tom. The mathematics of relativity specifically predict that Harry would make the round-trip in thirty-two years. Thus, upon landing on Earth, Harry would be only fifty-two years old. The twins would no longer be twins! Relativistic effects would have taken their toll.

Had the velocity of travel been greater, say 80 percent of the velocity of light, the trip time, as measured by stationary observers back on Earth, would have taken even less than thirty-two years. In fact, Harry would return from such a voyage only thirty-eight years old, while Tom would now be fifty. Their age discrepancy would have increased from eight years to twelve years as the relative velocity increased from 60 percent to 80 percent of the velocity of light.

The faster we travel, then, the more we postpone aging (at least relative to those who stayed home). However, we *cannot return to the past*. We can *grow old more slowly than normal*, but we cannot become younger. In the strictly hypothetical case of travel at the velocity of light, we could postpone aging altogether —in effect, achieving immortality by remaining in the present forever.

The dilemma here is that Harry finds nothing wrong in his spaceship. Indeed, nothing is wrong. All the laws of physics are completely identical in every reference frame either on Earth or anywhere else in the Universe; that's the first tenet of relativity theory. The traveler, Harry, perceives everything to be normal, provided he confines his view to objects on board the spaceship. He does not experience a slower pulse; he does not measure his spaceship clock, mass, or size to be abnormal. These physical quantities change only from the viewpoint of observers outside the spaceship, ones that have some velocity relative to the space-ship. (The paradox—namely, why Tom cannot also claim a delay in aging relative to Harry's viewpoint—is resolved only by study-ing the general theory of relativity, for only in that more grandiose theory can we account for the decelerations and accelerations experienced by Harry while turning around halfway through his round-trip voyage.)

Here, then, is a method by which, in principle at least, we can leap into the future. Provided the velocity is great enough, travelers might return to Earth when all their friends are dead and buried. At velocities close to that of light, travelers could conceivably return to an Earth centuries or even millennia in the future. In practice, however, the velocities needed for such exciting prospects are far beyond those attainable with current technology. As we noted earlier in this chapter, the highest velocity of an Earth-launched spacecraft *(Pioneer 11)* is about 0.02 percent of the velocity of light. Even the U.S. astronauts, whose time dilated during their week-long trips to the Moon, aged a mere 0.005 second less than the rest of us who stayed behind on Earth—a theoretical statement that is quite impossible to verify. So our civilization is still far from achieving appreciable relativistic time travel into the future. Such powered travel may never become a practical reality, because of the accompanying growth of the spaceship's mass, which we also discussed earlier.

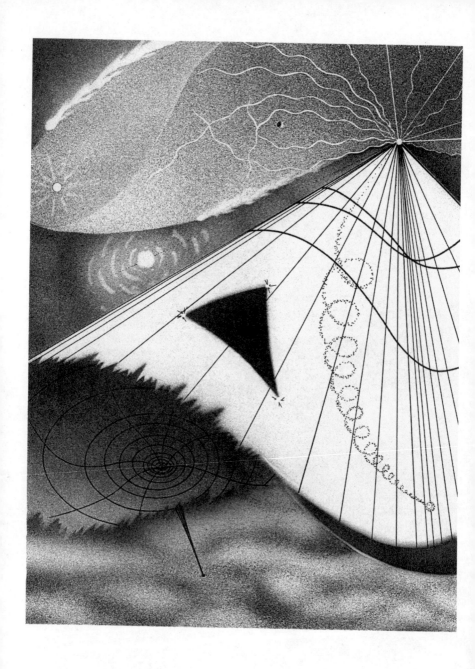

WARPED SPACETIME

SIX

The General Theory of Relativity

ALL THE previously described bizarre changes in length, mass, and time result directly from the theory of special relativity. We say "special" because, in all of the above analysis, the relative velocity between different reference frames was assumed to be constant. Einstein found it easier to examine first the consequences of uniform motion. A decade later (in 1915), in a work of stupendous originality, he relaxed that restriction by examining the consequences of variable motion, thereby making relativity less special—that is, more general.

Relativity for any velocity, constant or variable, is known as the general theory of relativity. In this intellectually broader form, the basic tenets of relativity remain essentially the same: first, the laws of nature are identical for all systems, regardless of their state of motion; and second, time is the fourth dimension. But we now consider spacetime reference frames traveling with nonconstant velocities—that is, accelerated reference frames.

Avoiding the (admittedly advanced) mathematics of general relativity, we can qualitatively explore, often only by analogy, the salient features of Einstein's greatest scientific contribution. One of the cornerstones of the theory can be appreciated by considering the following *Gedanken* (thought) experiment. Suppose we are in an elevator having no windows. When the elevator begins rising, we feel the floor pushing, especially on our feet. It is easy to attribute this pushing sensation to the upward acceleration of the elevator. Now, imagine that such a windowless elevator exists in outer space far from Earth. Normally, we would experience the weightlessness made familiar by watching the astronauts floating around where there are no net forces. But if we *did* experience a sensation of pushing on our bodies, we could draw one of two conclusions. We could argue that the elevator is accelerating upward in the absence of gravity, thus pinning us to the floor. Or we could maintain that the elevator is at rest in the presence of gravity, which is pulling us from below. In the first case, the rapid acceleration causes the sensation of pushing (or pulling); in the second, it is gravity that actually does the pulling. Without looking at objects outside the hypothetical elevator, we would have no way to determine which of these explanations is correct. In either case, pendulum clocks swing normally, released stones fall just as Galileo taught us, water pours from a glass in the customary fashion, and so on. Once we did build a "window" and look out, we would have no trouble establishing whether the elevator is really at rest or really accelerating. *Relative* to the Universe outside the elevator, we could easily assess the actual status of that elevator. Once again, as Einstein stressed by this hypothetical case, absolute motion does not exist.

The important point is this: the effect of gravity on an object and the effect of acceleration (inertia) on an object are indistinguishable. Expressed alternatively, a uniformly accelerated reference frame duplicates the behavior of a uniform gravitational

FIGURE 17 A windowless elevator accelerating through empty space in the absence of gravity *(top)* is indistinguishable from one at rest in the presence of gravity *(bottom)*. This is a statement of the principle of equivalence, which is at the heart of the general theory of relativity. Accordingly, Einstein suggested that the notion of acceleration is equivalent to the notion of gravity; what's more, he maintained that gravity, as an invisible force that pulls apples and other things to the ground, does not exist.

field. Scientists call this keystone of general relativity the "principle of equivalence": gravity and the acceleration of objects through spacetime can be viewed as conceptually and (almost) mathematically equivalent. Consequently, Einstein postulated as unnecessary the Newtonian idea of gravity as a force that attracts things. Not only is that idea illusory, but Newton's theory of gravitation is also today known to be less accurate than Einstein's. Let us briefly see how the concept of acceleration can replace the commonsense idea of gravitation.

General relativity predicts that mass changes the nature of spacetime. Bypassing the details (for they are mathematically formidable), we find that matter effectively shapes the geometry of spacetime. Put another way, mass is said to "curve" or "warp" spacetime.

Ordinary Euclidean geometry—the type learned in high school—holds valid when the extent of curvature is zero, that is, when spacetime is flat. Even when spacetime is only slightly curved, Euclidean geometry of flat space is approximately correct. At any one location on Earth's surface, for instance, an architect can design a building, or a contractor can build one, using procedures laid down twenty-five centuries ago by the Greek mathematician Euclid. However, though terrestrially familiar flat-space geometry is used every day by many workers, it is not absolutely correct. It can't be correct. Planet Earth, after all, is not flat; it's curved. On the surface of a sphere, flat Euclidean geometry works satisfactorily at any given (small) locality, but that's because we cannot clearly perceive our planet's curvature from any single location on its surface. Once the curvature of Earth becomes discernible, as it does in intercontinental aircraft or shipboard navigation, for example, a more sophisticated geometry must be used—a curved-space geometry.

And so it is at selected locations in the Universe; in the absence of matter, the curvature of spacetime is zero, the appropriate geometry is flat, and objects move undeflected in straight lines. Newtonian dynamics and Euclidean geometry are quite satisfactory, for all practical purposes, wherever spacetime is unappreciably curved. To be sure, flat space is not a hypothetical situation, since, far beyond the galaxy clusters, little or no matter presumably exists, thus ensuring only slight spacetime curvature.

On the other hand, the geometry of spacetime is strongly warped near massive objects. The object itself or the surface of the object is not warped, just the near-void of spacetime in which the object is embedded. The larger the amount of matter at any given location, the larger the extent of curvature or warp of spacetime at that location. Furthermore, the extent of warp grows progressively weaker at greater distances from a massive object. As with gravity, both the amount of matter and the distance from that matter specify the magnitude of spacetime curvature. But, since this view of warped spacetime is more accurate than the conventional view of gravity, to be precisely correct in all possible situations, the worldview of Newton and Euclid must be replaced by that of Einstein and other modern geometers.

Yet, surely you ask, how can a curve replace a force? The answer is that the topography of spacetime influences celestial travelers in their choice of routes much as Newton imagined gravity to hold an object in its path. Just as a pinball cannot traverse a straight path once shot along the side of a bowl, so the shape of space causes objects to follow curved paths (called geodesics). And any object whose motion changes direction, even though its speed may remain steady, is said to be accelerated. For example, Earth accelerates while orbiting the Sun—not because of gravity, as Newton maintained, but because of the curvature of spacetime, as Einstein preferred.

To see this, consider an analogy—not an example, an analogy. Imagine a pool table with a playing surface made of a thin rubber sheet rather than the usual hard felt. Such a rubber sheet would

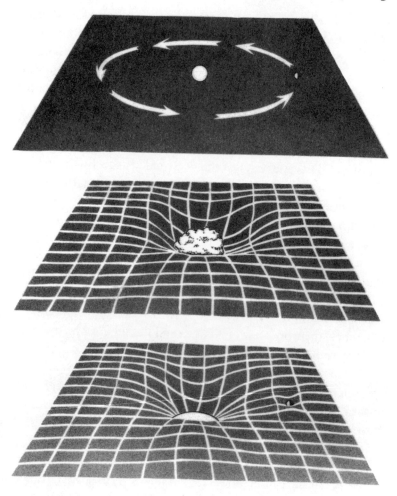

FIGURE 18 The "fabric" of spacetime can be visualized to be curved near a massive star *(bottom)* in much the same way that a rubber sheet distorts when a heavy rock is placed on it *(middle)*. The rock and the star themselves are represented by the central unshaded area in each frame. The response of a billiard ball to the rock's dimple in the (hypothetically frictionless) rubber sheet, or of the Earth to the Sun's warp in (real) spacetime, mimics the conventional view of our planet orbiting the Sun under the commonsense influence of gravity *(top)*.

become distorted if a large weight were placed on it. A heavy rock, for instance, would cause the sheet to sag or warp. The otherwise flat rubber sheet would become curved, especially near the rock. The heavier the rock, the greater the curvature. Trying to play pool, we would quickly find that balls passing near the rock are deflected by the curvature of the tabletop.

In much the same way, radiation and material objects are deflected by the curvature of spacetime near massive objects. For example, Earth is deflected from a straight-line path by the slight spacetime curvature created by our Sun. The extent of the deflection is large enough to cause our planet to circle repeatedly, or orbit, the Sun. Similarly, the Moon or a baseball responds to the spacetime curvature created by Earth, and they, too, move along curved paths. The deflection of the Moon is not too large, causing our neighbor to orbit Earth endlessly. The deflection of a small baseball is much larger, causing it to return to Earth's surface.

The commonsense notion of gravity, then, is just a convenient word for the natural behavior of objects responding to the curvature of spacetime. Accordingly, we can use a knowledge of spacetime to predict the motions of objects through space and time. More appropriately, we can turn the problem around: by studying the accelerated motion of any object, we can learn something about the geometry of spacetime near that object.

And so it is with the whole Universe. When we seek the size, shape, and structure of the entire Universe—the biggest picture of all—we need to take account of the net effect of spacetime curvature caused by each and every massive object in the cosmos. By studying the accelerated motions of various parcels of matter in the Universe, we can learn much about the net curvature of the whole Universe. We shall return to these ideas and do just that in chapter 11.

NAVSTAR, a "testbed" for both special and general relativity

SEVEN

Relativistic Tests

OW CAN WE be sure that the strange effects noted in earlier chapters really do occur when the relative velocity between observers becomes comparable to that of light? Likewise, how do we know for certain that mass curves spacetime? Before proceeding with further discussions of relativity and its applications, we need to examine the validity of the special and general theories of Albert Einstein. Fortunately, twentieth-century technology provides several tests of relativity. And the result of each test agrees almost precisely with the predictions of Einstein's theories.

The effects of special relativity, bizarre though they seem, are not idle predictions of an untested theory. Although we can hardly observe spacecraft having speeds close to that of the velocity of light, our civilization has built devices—called high-energy accelerators—capable of boosting subatomic particles to fantastically high velocities. In these giant laboratory machines, charged and magnetically guided particles whirl around and around, all the while repeatedly gaining energy and thus speed. Some of the lightest particles—for instance, electrons—can achieve velocities of approximately 0.999999999992 times the velocity of light. You can't get much closer to the velocity of light without actually achieving it. Accordingly, from the viewpoint of a stationary observer outside the accelerator laboratory, the relative velocity and hence the relativistic factor are large. In fact, direct observa-

FIGURE 19 This circular accelerator, at the Conseil Européen pour la Recherche Nucléaire (CERN), near Geneva, is used to guide and energize subatomic particles within a huge (several-kilometer diameter) underground ring. Some elementary particles can be boosted in the ring to velocities nearly equal to that of light itself. Though the particles cannot be seen directly, collisions among them yield debris that can be used to infer some properties of those particles. The philosophy of approach can be likened to smashing together two watches and then studying the scattered remains of springs, cogs, and gears (or, nowadays, silicon chips, integrated circuits, and liquid crystal displays) in order to determine how watches work!

tions prove that the electron's mass does indeed increase. For the above-noted velocity, electrons have been measured to be nearly 100,000 times more massive than their normal mass when at rest. This value agrees very closely with the theoretical predictions of relativity theory. Actually, the world's largest accelerators are designed and built with relativity in mind; if the theory were incorrect, the laboratory accelerators would not operate properly.

(Incidentally, the electron's "weight" gain is not a violation of any physical law, since its increased mass results from the electrical energy injected into the apparatus, again in accord with the basic equation $E = mc^2$, as we noted in chapter 5. This case also serves to point up why no piece of matter can "crash the light barrier" or even achieve the velocity of light itself; as particles accelerate to ever-higher velocities, they practically stop gaining in speed and instead pile on mass, which in turn makes them progressively harder to accelerate further.)

In a related experiment, a short-lived elementary particle is known to have its lifetime prolonged when traveling rapidly. These particles, called muons, are produced when cosmic-ray protons strike air molecules in the upper atmosphere of our planet. Normally (at rest), they persist for about two millionths of a second before decaying into energy and other, less massive particles; hence none of the muons should reach the ground. However, apparatus placed at sea level and atop Mount Washington, New Hampshire, was able to detect the fast-moving muons and to measure a significantly longer lifetime for them; relative to our fixed (Earth-based) reference frame, their "natural clocks" had apparently slowed down. In fact, while traveling at 99.4 percent of the velocity of light, muons endure for about sixteen microseconds, an eightfold increase in their usual lifetime. This time extension matches to within 1 percent that predicted by Einstein's theory of special relativity.

Even greater accuracy has been recently achieved with muons artificially generated in the CERN apparatus shown in figure 19.

Constrained by magnets, particles traveling at 99.94 percent of the velocity of light had their life expectancy dilated by a factor of twenty-nine. This experimental result came to within 0.1 percent of that predicted by Einstein's theory of time extension; it is the most accurate confirmation to date of special relativity and of the fantastic implications for "twin spaceflight" discussed toward the end of chapter 5.

Over the years, scientists have designed several ingenious tests of general relativity theory. The results of all these agree to a high degree of accuracy with the predictions of Einstein's theory. There are three principal tests.

The first concerns the orbital precession of the planet Mercury. As Mercury wheels around the Sun, its elliptical orbit precesses slightly. In a sense, its orbit wobbles like a top (though ever so slowly), advancing in the direction of its revolution around the Sun. Since about 200,000 Earth years are needed for Mercury's orbit to precess all the way around 360°, we can alternatively state that Mercury's precession amounts to nearly 600 arc seconds every Earth century. This is an observational fact, measured by astronomers long before the development of relativity theory. By summing the gravitational influences (perturbations) exerted by the other planets, Newton's theory of gravity can explain almost all this precession. But not entirely. It cannot account for 43 of the 600 arc seconds of precession per century. This is admittedly a small discrepancy, for an arc second is only $\frac{1}{3600}$ of an arc degree; in fact, the extra 43 arc seconds would grant Mercury, which normally revolves about the Sun every eighty-eight days, one extra revolution every few million years. But it's a discrepancy nonetheless. And it helped prove that Newton's ideas are not quite correct.

Using his relativity theory, Einstein was able to explain mathematically and precisely Mercury's orbital peculiarity. General relativity demands that Mercury's orbit precess ever so slightly

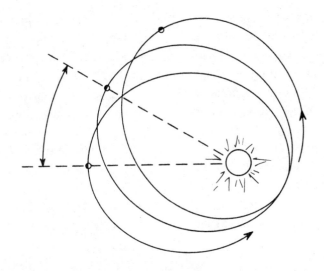

FIGURE 20 The orbits of all the planets precess over the course of time. Mercury's precession, greatly exaggerated here, is the largest of all the objects in the Solar System since this innermost planet is most affected by the Sun's spacetime curvature. (Only after about 17,000 Earth years would Mercury's orbit actually precess as much as the dashed lines indicate.)

more than predicted by Newton's theory because of Mercury's response to the way in which the Sun has warped spacetime. Although subtle, the added Einsteinian precession is in this case experimentally verified, thus favoring relativity theory.

Incidentally, all the planetary orbits are expected to experience some degree of precession, but, because the other planets are farther from the Sun, the extent of spacetime's curvature is less than for Mercury. Consequently, the amount of precession is also less for the other planets—so small, in fact, as to be currently unmeasurable.

A second, even more impressive test of general relativity concerns the gravitational bending of light, a phenomenon of which scientists before Einstein had scarcely dreamed. Relativity theory predicts that light, or any type of radiation, bends while passing through warped spacetime near a massive object. This bending, expected to be especially prominent as the light of some background star is deflected while grazing the edge of the Sun, occurs because radiation is composed of massless elementary particles called photons, each nonetheless having an *effective* mass of E/c^2 in accord with special relativity, as was discussed earlier. Each photon emitted by a distant star is therefore affected ever so slightly by the Sun. By altering their path in this way, the photons are actually traveling along the shortest route in curved spacetime.

So, when a remote star is nearly eclipsed by our Sun, the bending of light makes that star appear displaced from its normal

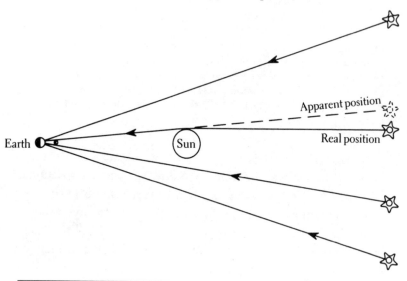

FIGURE 21 Radiation bends while grazing the edge of our Sun. The closer the observed light from a distant star gets to the Sun, the greater the difference between the real *(solid line)* and apparent *(dashed line)* positions of that star. The effect, as drawn here, is greatly exaggerated for clarity; the maximum bending of light passing the very limb of the Sun is actually less than two arc seconds, which is equivalent to the minute angle subtended by an American quarter viewed from a distance of some three miles. The light from other stars, whose rays do not closely bypass the Sun, is not measurably bent.

position among the other stellar members of its usual constellation. Light rays from the other stars, not aligned with the solar edge, pass through regions of lesser spacetime curvature near the Sun and hence are unappreciably bent on their way toward Earth. Even the change in the path of the starlight that actually grazes the limb of the Sun is very slight, amounting to approximately 1.75 arc seconds. (Though Newton was ignorant of this effect, a reanalysis of his worldview suggests that Newtonian gravity would also bend starlight, but by about half the amount predicted by Einstein's novel notions of spacetime curvature.) This minute angular change is actually measurable since the positions of most stars can be observed nearly a hundred times more accurately than an arc second.

Still, the measurement is an not easy one. After all, it is usually impossible to see any stars so close to our Sun, given its overwhelming brightness. However, with some effort, the theory can be checked during a total solar eclipse as the daytime skies darken while the Moon momentarily blocks the Sun. Indeed, ever since the British astronomer Arthur Eddington led an expedition to Africa in 1919, scientists have mounted relativity experiments to verify that light beams really do bend near massive objects. Furthermore, dozens of additional experiments have been performed in recent years, without the aid of an eclipse, by using cosmic radio sources whose radiation grazes the Sun. Other tests have involved spacecraft orbiting the Sun and exchanging radio signals when they were positioned so that the waves would pass close to the Sun, thus providing accurate measures of the time lag caused by our Sun's gravitational field (or spacetime curvature). Still other tests have used Earth-launched radar signals (and their returned echos) that grazed the Sun shortly before and after bouncing off the interior planets, Mercury and Venus, then on the far side of the Sun. The results of these optical and radio experiments almost invariably substantiate the prediction of relativity; the gravitational bending measured during several of the most recent experiments averages 1.7 ± 0.1 arc seconds.

A third classical test of general relativity goes by the name of gravitational red shift. Photons of light or any other type of radiation must do some work while leaving the surface of a massive object. Accordingly, all photons lose some of their energy while striving to escape from an object's strong gravitational force —or, as an Einsteinian might say, while the photons struggle to make their way up out of the large spacetime curvature in the object's immediate vicinity. Gravity (or spacetime curvature) literally pulls (or affects) the photons, each of which has an equivalent mass of E/c^2. Since photon energy is proportional to the frequency of the radiation composing those photons (a fundamental physical phenomenon first deciphered also by Einstein—indeed, a study that won him the Nobel Prize), this loss of energy shifts the frequency of the escaping radiation toward lower values.

The name "red shift" commonly refers to the change of white light to red light. But gravitational red shift applies to invisible radiation as well, causing all types of radiation to shift toward lower frequencies. An external observer would then expect to detect the radiation to be red shifted by an amount depending on the mass of the emitting object. And since the frequency of radiation is lessened in a strong gravitational field (or in a region of extensive spacetime curvature), this is tantamount to stating that a clock present in such a region should run with a slightly slower rhythm than on Earth. Apparently, gravity affects time; it slows it down.

Recognize that gravitational red shift has nothing to do with the motion of the source of the escaping photons; its origin is distinct from that of the motion-induced Doppler effect, which also causes radiation to be shifted (as we discussed regarding the galaxies, in chapter 1). The cause of gravitational red shift depends largely on mass and more approximately on density.

Usually, gravitational red shift is not a large effect. Radiation

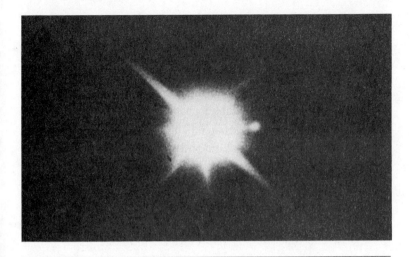

FIGURE 22 Gravitational red shift has been crudely observed in the radiation emitted by a handful of extraordinarily dense astronomical objects, such as the white-dwarf star Sirius B, shown here as the small dot of light to the right of center. The larger, more prominent object is its companion Sirius A (the Dog Star), the brightest star in the sky save the Sun. (The hexagonal shape of the image of Sirius A is not real; the "spikes" are artifacts caused by the support struts of the telescope.)

emitted by most stars, even including our Sun, is thought to be shifted, but not enough to be measured clearly. Extremely compact objects, with a gravitational force or, more correctly, a space-time curvature much greater than that of normal stars, should display an observable red shift of their emitted light. And some of them seem to, such as the compact white-dwarf star Sirius B, whose density is roughly a million times that of our Sun. While not unreliable, these astronomical observations nonetheless lack the accuracy to provide good confirmation of Einstein's theory.

Then how can we be sure that gravitational red shift really does occur as Einstein predicted? In particular, how can this shift be used as a third test of general relativity? Fortunately, an elegant experiment perfected at Harvard in 1965 has successfully confirmed the phenomenon of gravitational red shift. There, an apparatus was designed to measure the frequency change of a light beam while it travels from the bottom to the top of a seven-story tower. Since Earth's mass curves spacetime at the bottom of the shaft more than at the top, a small shift is expected in the frequency of radiation as the radiation moves from the basement to the roof. Unlike astronomical observations, this terrestrial experiment can be thoroughly controlled and the results recorded with high precision. Gravitational red shift has indeed been observed, and its measured value agrees almost exactly with the prediction of general relativity.

In a variation of this third test of relativity, and one that has very definite practical applications, a military navigational satellite (part of a constellation of spacecraft forming the U.S. Global Positioning System) has simultaneously confirmed the timing predictions of both special and general relativity. The satellite, *NAVSTAR 2*, orbits Earth at an altitude of some 20,000 kilometers and has on board a highly accurate atomic clock. First, because of the significant relative velocity between the satellite and its ground stations, special relativity predicts that the satellite-based clock will run slower (at a lower frequency) than ones located on the ground. Second, general relativity maintains that clocks run slower as they approach the center of a gravity field; this second effect thus clearly opposes the previous relative-velocity effect. The two effects would cancel each other for satellites having orbital radii of 1.5 times the Earth radius, but the *NAVSTAR* satellites have orbits of about 4.2 Earth radii, and hence their clocks run faster than Earth clocks. The total accumulated discrepancy due to both effects over the course of a day amounts to about 38.5 microseconds (faster than normal). If these effects were not compensated for (by adjusting the satellite clock), navigational ranging errors would average 11 kilometers per day. This twofold relativistic time dilation has been experimentally verified to within 1 percent of that predicted by Einstein, and thus the

FIGURE 23 Gravitational red shift has been carefully confirmed in a laboratory experiment where gamma-ray photons launched from iron nuclei at the bottom of a tower became redder by the time they reached a detector on the roof. The radiation effectively had to struggle to emerge from a region of considerable spacetime curvature, in the process doing some work and thus losing some of its energy.

corrected satellite system can provide navigational positioning with much-improved accuracy—in fact, on the order of a few meters.

Two other remarkable predictions of relativity theory might well have been recently confirmed. The first concerns an exotic variation of the bending of light described earlier in this chapter. In recent years, astronomers seem to have found several cases of what is called a "gravitational lens." Adjacent images of what at first sight appear to be identical twin quasars (distant, active galaxies) might be the result of the bending of a single quasar's radiation by some massive object (probably an ordinary galaxy) between us and the quasar. Even the elemental spectra of the two images are identical. As sketched in figure 24, this effect would mimic the way a glass lens of the proper shape can produce multiple images (or push the ends of an insect farther apart to make it look bigger).

Gravitational lensing derives directly from Einstein's theory, which predicts that light, or any type of radiation, bends while passing through warped spacetime near a massive object. That this is so follows from our above discussion in which we reasoned that photons possess an equivalent mass equal to E/c^2, and hence should be deflected ever so slightly while grazing the Sun. However, the Sun is hardly necessary for this rarity—indeed, more-massive and more-compact objects should be even more effective in deflecting radiation. Accordingly, as figure 24 depicts, a distant quasar's radiation could bend significantly while passing through an intervening galaxy (or rich cluster of galaxies). It is the intervening galaxy that acts as a gravitational "lens," which, if it has a strong enough gravitational force field, can form multiple images of a single celestial object.

In yet another potential verification of general relativity, astronomers might well have confirmed its prediction that a kind of radiative wave ought to be emitted by virtue of an object's mass.

To discern Einstein's sagacity here, recall that electromagnetic waves are a common, everyday phenomenon. Briefly, they are well-known cyclical disturbances that move through space, transporting energy from one place to another. Whether they take the

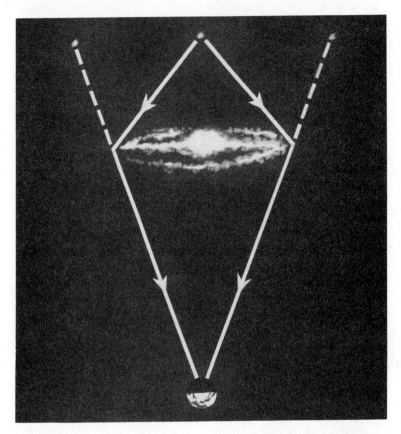

FIGURE 24 A massive object, such as a galaxy (drawn here as a large spiral), can act as an astronomical lens by bending radiation from some distant object (depicted here as a quasar directly behind the galaxy). Accordingly, false images can thereby be projected to each side of the intervening galaxy, as seen along the dashed lines from Earth, at the bottom of the figure. In this sketch, as in the closely allied figure 21, the effect of the gravitational lens is greatly exaggerated for clarity.

form of radio, infrared, light, ultraviolet, X-ray, or gamma-ray radiation, electromagnetic waves are essentially caused by changes in the strengths of electromagnetic force fields. Oscillations of charged electrons within a broadcasting tower, for example, generate speed-of-light electromagnetic waves, which are then received by "shaking" other electrical charges in the antennas on the roofs of our homes. Measurements of these waves agree perfectly well with the theory of electromagnetism that predicts them.

The modern theory of gravity—Einstein's theory of relativity—also predicts cyclical disturbances that move through space. A "gravity wave" is the gravitational analogue of an electromagnetic wave. Gravity waves, or gravitational radiation, result from changes in the strengths of gravitational force fields. In principle, every time an object of any mass accelerates, gravity waves should be emitted at the velocity of light. And, like electromagnetic waves, gravity waves should cause other masses to vibrate or shake in response to their passage. In practice, though, no one has ever unambiguously detected such gravity waves.

Although gravity waves are a possibility, given relativity theory, their existence is still not guaranteed. The formal mathematics of theoretical physics portends many things that seem to disagree with physical reality. The converse is also often true, at least for a while. For example, until rather recently, many scientists regarded the ideas of black holes and neutron stars as absurd. But, as we shall see in part IV, these and other exotic phenomena are now gaining acceptance, because observational evidence is hard to deny.

Gravity waves would not be easy to capture, even if we knew how to do it. Part of the trouble arises because theorists are still arguing about what kinds of astronomical objects should produce intense gravity waves. Leading candidates include (1) the merger of a binary-star system wherein two stars orbiting about one another spiral toward their center of mass, (2) the collapse of a star into a black hole (see chapter 16), and (3) the collision of two black holes. Each of these possibilities involves the accelerations of huge masses. In this way, the strength of the gravitational force field should change (in the latter two cases drastically), causing a gravity wave to propagate outward from the source. Other cosmic objects are also expected to emit gravity waves, but only changes in large amounts of mass will produce such waves of observable magnitude.

Interestingly enough, a steady change in the orbit of a recently discovered binary-star system is now being monitored for gravity waves. This galactic binary system is unique since it contains a spinning star that radiates bursts of rhythmic radio signals 17 times per second, much like flashes from a rapidly swiveling lighthouse beacon; such compact, pulsating stars are colloquially called pulsars. This "binary pulsar," named 1913+16 and about 15,000 light-years away, is also unique since the periodic Doppler shift of its radiation proves that the orbital system is slowly decaying; the system is precessing some 30,000 times faster than the advance of Mercury's orbit around the Sun. As the two objects in the binary system gradually spiral toward one another, their orbital energy transforms into gravitational energy. This energy should theoretically be radiated away as gravity waves. However, no such waves have yet been detected directly.

Astrophysicists need to develop experimental tools to detect gravity waves from a variety of cosmic objects. Radiation is energy, and energy is information. Gravity waves should therefore contain a great deal of information about the physical events in some of the most exotic regions of space. The discovery of gravity waves could herald a new age in astronomy, in much the same way that the discovery of invisible electromagnetic waves, unknown a century ago, revolutionized classical astronomy and led to modern astrophysics.

EIGHT

More about Curved Spacetime

HAVING NOTED in chapter 7 that, despite its rather peculiar consequences, relativity theory has survived many experimental tests, we illustrate here some additional features of spacetime curvature. Through further discussions, examples, and analogies, we can gain a better appreciation of this stunningly original concept of modern science.

Recall from chapter 6 that, by including the notion of gravity within the special theory of relativity, Einstein was able to develop a more general theory of relativity describing the motions of objects having any velocity, constant or otherwise. Furthermore, this broader theory stipulates that the response of objects to the curvature of spacetime virtually mimics the concept of gravity. The results of the two—gravitational force and space-

Einstein in his study in Berlin, 1919

time curvature—are nearly indistinguishable, for each depends similarly upon mass and distance.

Operationally, though, the two views differ: Newton's theory stipulates that all of space is filled with invisible gravitational fields extending from every massive object in the Universe; Einstein's theory maintains that the (equally invisible) "fabric" of spacetime is warped because of every massive object in the Universe. The effect of massive objects on the large-scale structure of the Universe can then be studied either in the Newtonian sense, by measuring the strength of their gravitational forces, or in the Einsteinian sense, by watching the behavior of other objects accelerating near them. The Newtonian and Einsteinian views predict virtually the same results in most cases. However, under exceptional circumstances—when objects have either very high velocities or very large masses—Newton's idea fails, and only Einstein's prevails. Or, stated in perhaps more sober terms, Einstein's notions prove to be a better approximation of reality than Newton's. At any rate, since some objects, such as distant galaxies and black holes, are suspected to have either high velocities or large masses (and sometimes both), we must use relativity theory in many cases if we are to obtain answers as correct as humanly possible.

To illustrate further the curvature of spacetime, ponder the following hypothetical case. Imagine three planets to be inhabited by equally advanced civilizations capable of launching rockets. Suppose we take Earth, Mars, and Jupiter. Of these, Mars is the least massive and Jupiter the most massive. Suppose, furthermore, that the inhabitants of each planet possess identical rocket technology. And, for discussion's sake, let us assume that these rockets can achieve only a fixed amount of thrust (energy) at launch, after which they glide freely through space.

When the rockets are launched from each of the three planets, the shapes of their trajectories differ. In the Newtonian view of space, the rocket paths are determined by the gravitational interaction between the rocket and each planet. In the Einsteinian view of spacetime, these paths are determined by the response of the rocket to the spacetime warp created by each planet.

Figure 25*(a)* depicts a possible path of a rocket launched from the most massive planet, Jupiter. As is shown, the initial thrust was chosen in this case to be large enough to place the rocket into an elliptical orbit. Like gravity, whose strength decreases with increasing distance from a massive object, the curvature of spacetime is also greater close to the massive planet. The rocket accordingly speeds up (or accelerates) when close by and slows down (or decelerates) when far away. General relativity thus agrees with the Keplerian laws of planetary motion, stipulating that rockets move faster (accelerate) near massive objects, owing to the greater degree of spacetime curvature there.

An ellipse, a "closed" geometrical path, is only one possible type of motion. It is the trajectory of minimum energy, so labeled because a rocket in such an orbit does not have enough energy to escape the planet's influence.

Rockets can have other paths as well. Figure 25*(b)* shows a typical path taken by an identical rocket after launch from the less massive planet Earth. The same thrust used to launch the Jupiter rocket into an elliptical orbit is now great enough to propel the rocket entirely away from Earth. Less energy is used in the launch from Earth than in that from Jupiter, and thus more energy is imparted to the motion (that is, kinetic energy) of the Earth rocket. The rocket escapes the influence of Earth because, as a Newtonian classicist would say, Earth has less gravitational pull than Jupiter. Alternatively, Einsteinian relativists maintain that such a rocket escapes from Earth because our planet warps spacetime much less than does Jupiter. The two views—Newtonian and Einsteinian—predict nearly identical paths for the rocket as it recedes toward regions of spacetime progressively less curved by planet Earth.

The resultant path shown in figure 25*(b)* is called a parabolic trajectory. This is the type of flight path taken by human-made spacecraft that have been probing the other planets of our Solar System in recent years; the trajectory exemplifies a more "open" geometry. An object traveling along a parabolic path has more energy than one on an elliptical trek, either because the initial thrust needed to achieve a parabolic trajectory was large or because the mass of the parent object from which the launch was made is relatively small. In the example cited here, the rockets were assumed to be identical, so the increased energy of the parabolic case results from the comparatively small mass of Earth.

Even while receding far from its parent planet, a rocket is still affected by the pull of gravity or the warp of spacetime created by the mass of that planet. Although large only in the immediate vicinity of the planet itself, Earth's influence over the rocket never diminishes to zero. Mathematical analyses predict that, in the idealized absence of all other astronomical objects, such a parabolically launched rocket should approach infinity—and will have a velocity of zero when it gets there. (Theoretically, we say that the rocket will just barely make it, and then stop! But since nothing can ever really reach infinity, this is tantamount to saying that the rocket will continue to recede forever.)

The parabolic path contrasts slightly with another type of trajectory conceivably taken by an escaping rocket. Figure 25*(c)* shows this third geometrical path, also open in form and called a hyperbola. In our example, it is the path taken by a rocket launched from Mars, the least massive of the planets considered. The hyperbolic path closely mimics the parabolic path, though they differ a little in energy content. Since less energy is used to launch identical rockets from Mars than from Earth, the receding Martian rocket has more energy of motion. And with the least amount of gravity or spacetime warp near Mars, the hyperbolically launched Martian rocket approaches infinity with little dif-

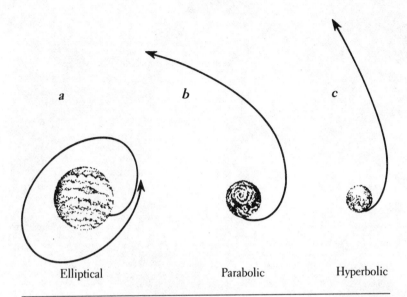

a   *b*   *c*

Elliptical   Parabolic   Hyperbolic

FIGURE 25 These are the resultant paths of identical rockets launched from three different planets—a very massive Jupiter *(left)*, a moderately massive Earth *(center)*, and a least massive Mars *(right)*. The trajectory of a rocket in elliptical orbit represents a "closed" geometry, whereas parabolic and hyperbolic trajectories are "open" geometries.

ficulty. (Having more energy than the parabolically moving rocket, the hyperbolic counterpart will theoretically reach infinity with some finite velocity larger than zero. Thereafter, the rocket will not stop, so claims the mathematics; it will continue moving beyond! The academic language of the mathematician notwithstanding, we realize that in actuality no object can reach infinity; parabolically and hyperbolically moving objects are destined to recede *toward* infinity indefinitely.)

The above three cases conveniently describe the motion of any object in terms of its energy content and of its response to spacetime curvature. These will be useful analogies when we consider, in chapter 11, the application of relativity to cosmology, for then the "object" will be the entire Universe itself.

The man in his study, 1947

NINE

Einstein, the Man

ALBERT EINSTEIN was not only a genius but also one of the most profound thinkers of all time. In addition to conceiving the theory of relativity, he contributed greatly to many other areas of science. Yet he was more than just a scientist. A man of unpretentious disposition, he was also a pacifist with a deep concern for the welfare of his fellow human beings. In 1931, during a lecture in California, he had this to say:

Concern for man himself and his fate must always form the chief interest of all technical endeavors, concern for the great unsolved problems of the organization of labor and the distribution of goods—in order that the creation of our mind shall be a blessing and not a curse to mankind. Never forget this in the midst of your diagrams and equations.

In many ways, Einstein thought of himself as a philosopher more than a scientist. Indeed, he emulated the Greek philosophers, attempting to account for natural phenomena as the basis of logical deductions instead of experimentation. Whereas the ancients failed, Einstein succeeded largely because he ingeniously utilized the powerful analytic tools of mathematics developed since the time of Aristotle.

Born in 1879 near Munich, into an open and tolerant family, Einstein was not a particularly bright student—at least not in formal book learning. At one and the same time, he was slow to talk (finally speaking at age three and becoming reasonably fluent in German at age nine) yet devoured mathematics texts crammed with equations (for which he had a natural affinity). Declaring that it was stupid to memorize things, he was expelled as a teenager from the austere, Prussian-disciplined Luitpold Gymnasium for being "a disturbance for the class and a negative influence on the other pupils"; in fact, he never did receive a high-school diploma. His distaste for the rigidity of academics continued throughout his university years (having been admitted to the Federal Polytechnic at Zurich not by passing a general entrance examination, which he failed, but on the strength of his exceptional prowess in mathematics):

Physics too [as taught in the classroom] was split into special fields each of which could engulf a short life's work without ever satisfying the hunger for deeper knowledge. For the examinations, one had to stuff oneself with all this rubbish, whether one wanted to or not. This conclusion had such a negative effect on me that after my finals, the consideration of any scientific problems was distasteful to me for a whole year.

Clearly, the interests of the gentle and taciturn Einstein were beyond (or at least outside of) the staid environment of the classroom. Interestingly enough, by his own admission, it was books of popularized science that awakened in him a passion for knowledge.

Einstein's independent character continued to plague him, and he was unable to secure a university position at the completion of his studies in 1900. His views were too original; his unorthodox spirit simply terrified the professors. To earn a living, Einstein was forced to accept a regular job as an examiner in the

Patent Office in Bern, Switzerland. Across his desk came a great variety of patent applications from which, he later confessed to his demanding boss, he learned much while striving to distill and describe in carefully chosen words the essence of each putative invention. More than from either his parents or his formal schooling, this is where Einstein gained a sense of what physics is and how it works. Hence his life-long fascination, despite his demonstrable theoretical bent, with the action of a compass and the subtleties of a gyroscope. John Archibald Wheeler, one of today's foremost relativists, summed it up poetically: "Einstein's sun was the great tradition of the great thinkers. His moon and stars were the happy mysteries hidden in things."

In 1903, Einstein married Mileva Marić, his former university companion, of Serbian origin. He had two sons by Mileva, Hans Albert (an engineer) and Eduard ("Tede," who died in 1965 in a psychiatric hospital), but the union was not happy and ended in 1919 with a quiet divorce. As alimony, Einstein promised to pay his ex-wife the interest on the 30,000 kronor awarded with the Nobel Prize—a curious detail, if only because at the time of his commitment he had not yet won the prize. (He took it a few years later, in 1921.)

Einstein's extraordinary ability to concentrate on problems with superhuman energy, to get to the crux of an issue and not let himself be distracted by details, is perhaps best exemplified during an eight-week interval in 1905. Therein, the twenty-six-year-old scientist gave the world three works that shook the foundations of science; more than any other group of papers, these publications mark the passage from classical to modern physics. These contributions addressed the quantum nature of light, the microscopic properties of particles in a gas, and the special theory of relativity.

Certain experiments conducted around 1900 had yielded very peculiar results concerning the way light interacts with matter. In what is now known as the photoelectric effect, when light shines on a metal surface, it knocks electrons out of it. But that's not all. As the intensity of the beam of light is weakened, the energy of the liberated electrons does not so diminish; rather, though fewer in number, the electrons continue to be freed from the metal with as much energy (that is, velocity) as before. This was (and still is) a preposterous result for those accustomed to regarding light as a smoothly undulating wave. But the experiment makes perfectly good sense if light, indeed every kind of radiation, consists of minute, invisible particles. This is precisely what Einstein declared, thus bolstering an earlier conjecture by the German physicist Max Planck that light is sometimes manifest as discrete bundles (photons); it also provided the essential clue that the Danish theorist Niels Bohr needed to launch a revolution in our understanding of the microscopic domain—a radical departure from past thinking that paralleled the equally significant revolution that Einstein himself would later begin in regard to the macroscopic domain.

In his second major advance of 1905, Einstein reduced to a clear formula the disorderly movement of microscopic particles. Nearly a century before, the Scottish botanist Robert Brown had demonstrated by mixing pollen dust in water that the component particles of the water were subject to incessant motions; the specks of dust moved in irregular zigzags, without the intervention of an external influence such as tidal or electric currents or any other action. Knowing how (the then hypothetical) molecules should behave in a gas, Einstein was able to prove that the so-called Brownian motion of the dust corresponded precisely. This work revealed for the first time the reality of molecules, whose existence had previously been a matter of controversy.

And in his third contribution of that momentous year, Einstein published his most revolutionary article to date, "On the Electrodynamics of Moving Bodies"—the relativity paper. Needless to say, the subject of this work caused quite a stir, not least because its consequences clashed with conservative physics and "common sense"; in a few pages, Einstein negated both the ether and the concept of absolute space, maintained that time did not pass uniformly as had been accepted for centuries, implied that matter could be directly transformed into energy and vice versa, and predicted all manner of baffling alterations regarding the relative measurements of length, mass, and time. What's more, as the historian of science Arthur Miller has noted,

Its title had little to do with most of its content; it had no citations to current literature; a significant portion of its first half seemed to be philosophical banter on the nature of certain basic physical concepts taken for granted by everyone; the only experiment explicitly discussed could be explained adequately using [then] current physical theory and was not considered to be of fundamental importance.

Yet, the still little-known author was able to derive exactly a result that had previously required several drastic approximations. And some key scientists quickly recognized its brilliance. The physicist Max Born wrote just a few months after the paper's appearance, "In my view one of the most remarkable works in the whole of scientific literature has been published this year in *Annalen der Physik*, on the subject of relativity."

Thereafter, Einstein's status in the scientific community skyrocketed, though he had not yet become a world celebrity. By 1909 he was awarded the title "Professor extraordinarius" at the Zurich Polytechnic and in 1910 a Prague University professorship by decree of Emperor Franz Josef of Austria; in 1913 he returned to his native country at the request of Emperor Wilhelm II of Germany, to assume a prominent professorship at Berlin University as well as the directorship of the country's research organization for theoretical physics.

As a staunch pacifist on the eve of war in the world's most militaristic nation, even the naturally unorthodox Einstein was ill at ease. His fame and his qualification as a neutral saved him from popular reprisals and academic ostracism, but he found himself rather quickly isolated. Despite his standing in the scientific community, he was all but ignored by colleagues whenever he took his chair at the Prussian Academy of Sciences. "Will future generations," he wrote to a friend, "really be able to glorify our Europe where three centuries of the most intensive cultural work have ended in nothing more than a change from religious to nationalist mania. Even the scholars here and in France behave as though their brains have been amputated."

The narrow-mindedness of his academic colleagues, both in science and in society, only served to amplify his educational ennui, resulting in a kind of professionalization of his earlier distaste for institutional conservatism; he sought to withdraw from the pettiness and squabbles of his university department. What most irritated the great relativist was later recalled by a student of Einstein's, Philipp Frank:

The daily life of a university often leads to discussion. [Academics] worry about the frequency at which works should be published, about colleagues who have or have not published anything, about a colleague who too often or perhaps not often enough quotes another colleague or who has failed intentionally or unintentionally to quote someone else. There are debates on the merits of certain professors, on the honors they hold or have not been awarded by their university or other universities, on the academies to which they have or have not been elected. The conversation turns to the number of students whom the professors have been able to provide with a post; to students and masters who have been capable of preventing them from finding posts; to the question of knowing if they have any influence on their superiors or if they are capable of obtaining money from the authorities for their department.

Despite his troublesome professorial life, Einstein managed to derive considerable solace from a rather peaceful home life. Living with relatives in Berlin, he took a liking to his uncle's amiable daughter, Fräulein Elsa, and eventually married her. Elsa, in turn, was able to detach him from the tiresome details of everyday routine, preserving his privacy and allowing him to work in great, uninterrupted streams. It was in this almost monastic environment that Einstein undertook the bulk of his greatest work—the all-embracing general theory of relativity, which brought him world fame.

That fame came rather suddenly when, on 29 May 1919—the day some say the modern world began—photographs of a solar eclipse, taken on the island of Principe, off West Africa, and at Sobral, in Brazil, confirmed the essence of a new theory of the Universe. No scientific test, before or after, captured so many global headlines. The actual announcement of light's bending around the Sun came in September when, at a packed meeting of the Royal Astronomical Society in London, Eddington presented the results of his expedition. To the philosopher Alfred North Whitehead, who was present, the meeting resembled a Greek drama:

We were the chorus commenting on the decree of destiny as disclosed in the development of a supreme incident. There was dramatic quality in the very staging: the traditional ceremonial, and in the background the picture of Newton to remind us that the greatest of scientific generalizations was now, after more than two centuries, to receive its first modification . . . a great adventure in thought had at last come home to shore.

Einstein had become the subject of worldwide conversation. He was in great demand, mobbed wherever he went, and became the living archetype of the abstract natural philosopher. Yet most people, then and later, never understood Einstein's work. The familiar world of Newtonian straight lines and Euclidean right angles was (and is) just too comfortable to abandon. To both persons in the street and professional nonspecialists, relativity theory has never become more than a vague source of unease. Even so, Einstein's research has changed our ideas, our perspective on the world. Much as Galilean empiricism cleared the way for the scientific and industrial revolutions, and much as Newtonian physics laid the basis for the eighteenth-century Enlightenment, Einsteinian relativity helped cut society from its moorings of common sense and human intuition. All the more reason, perhaps, that Einstein often stressed the need to communicate popularized but good, solid science to inquiring laypersons:

It is of great importance that the general public be given an opportunity to experience—consciously and intelligently—the efforts and results of scientific research. It is not sufficient that each result be taken up, elaborated, and applied by a few specialists in the field. Restricting the body of knowledge to a small group deadens the philosophical spirit of a people and leads to spiritual poverty.

For the remaining three and a half decades of his life (much of which was spent at the Institute for Advanced Study in Princeton, New Jersey), Einstein labored obstinately to achieve a complete unification of all of nature's forces. Early in the century, when only gravity and electromagnetic forces were known, he strove to find equations that would allow him to conclude that electric charges curve space as do masses; in this way, other charges passing nearby would deviate in their paths, thus giving the impression of a force. The "force of electromagnetism," like the "force of gravity," could then be postulated as fiction. Alas, he never succeeded in synthesizing these two forces, and the discovery toward midcentury of two more (the strong force that binds nuclei and the weak force that governs the decay of radioactive atoms) merely complicated his dream of unifying the most fundamental aspects of physics.

In the eyes of many researchers (then and now), Einstein never could have succeeded in this grand quest, for he thoroughly rejected the essence of the other great revolution in early-twentieth-century science; he refused to accept the statistical or chancy nature of the microscopic world, which is the hallmark of quantum physics. He was at one and the same time a founder of the quantum interpretation of the microscopic domain and a strict determinist unwilling to concede the probabilistic nature of that domain. In a letter to a friend, Einstein put it succinctly: "God does not play dice."

Interestingly enough, some three decades after Einstein's death and nearly three-quarters of a century after he first began to pursue his "grand-unified theory," we now feel poised on the threshold of achieving just such a broad synthesis of all the known forces of nature; these modern efforts will be discussed in part IV.

Given his world status, Albert Einstein was naturally asked to comment on all manner of global issues, from religion and morality to society and politics. This played well on his wide cultural interests, and, despite his early reluctance to entertain distractions from his research, later in life he had much to say and write about a highly diverse set of subjects. Here is a brief sampling of the man's opinions on a few representative issues; for more of his views, the reader is referred to some of his original writings listed in the bibliography at the end of this book.

Einstein had much to contribute on morality, including the following:

Needless to say, one is glad that religion strives to work for the realization of the moral principle. Yet the moral imperative is not a matter for church and religion alone, but the most precious traditional possession of all mankind. Consider from this standpoint the position of the Press, or of the schools with their competitive method! Everything is dominated by the cult of efficiency and of success and not by the value of things and men in relation to the moral ends of human society. To that must be added the moral deterioration resulting from a ruthless economic struggle. The deliberate nurturing of the moral sense also outside the religious sphere, however, should help also in this, to lead men to look upon social problems as so many opportunities for joyous service towards a better life. For looked at from a simple human point of view, moral conduct does not mean merely a stern demand to renounce some of the desired joys of life, but rather a sociable interest in a happier lot for all men.

In addition to the telling quotation that began this chapter, here he is again regarding adequate balance:

It is true that convictions can best be supported with experience and clear thinking. On this point one must agree unreservedly with the extreme rationalist. The weak point of his conception is, however, this, that those convictions which are necessary and determinant for our conduct and judgments, cannot be found solely along this solid scientific way.

For the scientific method can teach us nothing else beyond how facts are related to, and conditioned by, each other. The aspiration toward such objective knowledge belongs to the highest of which man is capable, and you will certainly not suspect me of wishing to belittle the achievements and the heroic efforts of man in this sphere. Yet it is equally clear that knowledge of what *is* does not open the door directly to what *should be*. One can have the clearest and most complete knowledge of what *is*, and yet not be able to deduct from that what should be the *goal* of our human aspirations. . . . The knowledge of truth as such is wonderful, but it is so little capable of acting as a guide that it cannot prove even the justification and the value of the aspiration towards that very knowledge of truth.

To Einstein, religion meant this:

The highest principles for our aspirations and judgments are given to us in the Jewish-Christian religious tradition. It is a very high goal. . . . If one were to take that goal out of its religious form and look merely at its purely human side, one might state it perhaps thus: free and responsible development of the individual, so that he may place his powers freely and gladly in the service of all mankind.

And when pressed further on religious inclinations, he made clear his undiluted pantheism:

I believe in Spinoza's God who reveals himself in the harmony of all that exists, but not in a God who concerns himself with the fate and actions of men.

On ethics, Einstein had much to say, of which this is a very small excerpt:

As far as I can see, there is one consideration which stands at the threshold of all moral teaching. If men as individuals surrender to the call of their elementary instincts, avoiding pain and seeking satisfaction only for their own selves, the result for them all taken together must be a state of insecurity, of fear, and of promiscuous misery. If, besides that, they use their intelligence from an individualist, i.e., a selfish standpoint, building up their life on the illusion of a happy unattached existence, things will be hardly better. In comparison with the other elementary instincts and impulses, the emotions of love, of pity and of friendship are too weak and too cramped to lead to a tolerable state of human society.

The solution of this problem, when freely considered, is simple enough, and it seems also to echo from the teachings of the wise men of the past always in the same strain: All men should let their conduct be guided by the same principles; and those principles should be such, that by following them there should accrue to all as great a measure as possible of security and satisfaction, and as small a measure as possible of suffering.

Throughout much of his later life, Einstein championed the idea of world government, though on this issue he despaired, for he often alienated both East and West alike:

Unfortunately, there are no indications that governments yet realize that the situation in which mankind finds itself makes the adoption of revolutionary measures a compelling necessity. Our situation is not comparable to anything in the past. It is impossible, therefore, to apply methods and measures which at an earlier age might have been sufficient. We must revolutionize our thinking, revolutionize our actions, and must have the courage to revolutionize relations among the nations of the world. Clichés of yesterday will no longer do today, and will, no doubt, be hopelessly out of date tomorrow. To bring this home to men all over the world is the most important and most fateful social function intellectuals have ever had to shoulder. Will they have enough courage to overcome their own national ties to the extent that is necessary to induce the people of the world to change their deep-rooted national traditions in a most radical fashion?

A tremendous effort is indispensable. If it fails now, the supranational organization will be built later, but then it will have to be built upon the ruins of a large part of the now existing world. Let us hope that the abolition of the existing international anarchy will not need to be brought by a self-inflicted world catastrophe the dimensions of which none of us can possibly imagine. The time is terribly short. We must act now if we are to act at all.

And, finally, here is Einstein on his own being:

For the most part I do the thing which my own nature drives me to do. It is embarrassing to earn so much respect and love for it. Arrows of hate have been shot at me too; but they never hit me, because somehow they belonged to another world, with which I have no connection whatsoever.

For many years before his death, Einstein lived the life of a recluse in Princeton, striving to achieve his unified field theory while occasionally speaking out against social injustice and for world government. He worked in his office to the end, aged rather suddenly in his last year or so, and died in mid-1955.

Of all the glowing testimonials offered Einstein in death, perhaps the most poignant was this one by the British author-scientist C. P. Snow:

Einstein was the most powerful mind of the twentieth century, and one of the most powerful that ever lived. He was more than that. He was a man of enormous weight of personality, and perhaps most of all, of moral stature. . . . I have met a number of people whom the world calls great; of these, he was by far, by an order of magnitude, the most impressive. He was—despite the warmth, the humanity, the touch of comedian—the most different from other men.

Albert Einstein in Berlin, 1920

With Elsa en route to the United States, 1921

With Max Planck in Berlin, 1929

Albert Michelson, Einstein, and Robert Millikan at Caltech, 1931

In Pasadena, 1931

Einstein lecturing in 1931

Einstein in Princeton with his longtime secretary, Helen Dukas

Einstein, Elsa, and Charlie Chaplin, Los Angeles, 1931

ON THE FACING PAGE
Above, Einstein with Harlow Shapley, 1935
Below, Einstein with the astrophysicist Donald Menzel and the mathematician
Garrett Birkhoff at the Harvard Observatory residence, 1935; the boy is
Carl Shapley

Sailing on Saranac Lake, 1936

ON THE FACING PAGE: Einstein in the mid-1950s

His office, as he left it at death

COSMOLOGY

Analogy with classical mechanics shows that the following is a way to complete the theory. One sets up as field equation

$$R_{ik} - \tfrac{1}{2}g_{ik}R = -T_{ik}$$

where R represents the scalar of Riemannian curvature, T_{ik} the energy tensor of the matter in phenomenological representation. The left side of the equation is chosen in such a manner that its divergence disappears identically. The resulting disappearance of the divergence of the right side produces the "equations of motion" of matter, in the form of partial differential equations for the case where T_{ik} introduces, for the description of the matter, only *four* further functions independent of each other (for instance, density, pressure, and velocity components, where there is between the latter an identity, and between pressure and density an equation of condition).

By this formulation one reduces the whole mechanics of gravitation to the solution of a single system of covariant partial differential equations. The theory avoids all internal discrepancies which we have charged against the basis of classical mechanics. It is sufficient —as far as we know—for the representation of the observed facts of celestial mechanics. But, it is similar to a building, one wing of which is made of fine marble (left part of the equation), but the other wing of which is built of low grade wood (right side of equation). The phenomenological representation of matter is, in fact, only a crude substitute for a representation which would correspond to all known properties of matter.

—A. Einstein, *Out of My Later Years*

Einstein addressing his colleagues, 1922

TEN

Cosmological Principles

B Y INFUSING the basic tenets of relativity throughout a full-blown (mathematical) development
of the theory, researchers have learned to map
the varied ways that matter warps spacetime.
This is the area where relativity becomes notoriously complex;
here, what we glean from their ponderous calculations can only
be appreciative. The results, in a nutshell, are the so-called Einsteinian field equations—a group of a dozen or so equations that
must be solved simultaneously to determine the grandest structure of space and time. These mathematical formulas specify the
curvature of spacetime either locally (in the vicinity of some
isolated object) or more generally (owing to matter present
throughout the Universe).

Although on the one hand these equations are extremely
formidable to solve quantitatively, on the other they embody
remarkable symmetry qualitatively. Much like works of art, they
often inspire a sense of wonder, a certain awe. The complexity
arises largely because, in addition to the field equations specifying
the geometry of the Universe, the relativist must also solve several

other formulas—called the geodesic equations—to determine the response of individual objects to the curvature of spacetime at any location in the Universe. In the case of the planets, that object might be a rocket like the one in chapter 8; in the case of the entire Universe, our view necessarily broadens and the object becomes, for example, a galaxy.

The study of these coupled sets of equations—the field equations specifying spacetime geometry and the geodesic equations describing spacetime dynamics—has been given by Wheeler the tongue-twisting name of geometrodynamics. Let us now consider what I termed earlier "the broadest view of the biggest picture" and explore what geometrodynamics has to say about the Universe as a whole.

Because Einstein created general relativity, he was obviously better positioned than anyone else to use his equations to deduce the nature and structure of the cosmos. In 1917, his field equations predicted the net curvature of the whole Universe to be large. He found that the flat geometry of Euclid just would not suffice when one examined the bulk properties of the entire Universe. Unfortunately, Einstein's solution can be cast only in terms of the unimaginable, four-dimensional spacetime. It is fully imaginable mathematically, but quite unimaginable conceptually.

The essence of his solution can nonetheless be visualized by using another analogy. No one to my knowledge has ever built or articulated a viewable example of anything in four dimensions, so in this analogy we suppress one of those four dimensions. For the sake of argument, suppose we consolidate the three dimensions of space into only two dimensions, much as we compressed spatial dimensions in chapter 4. Then, with time as the remaining dimension, we can construct a three-dimensional analogue of Einstein's four-dimensional Universe.

Our analogue is a three-dimensional sphere, sometimes colloquially termed "Einstein's curveball." Here, all of space is spread *on the surface* of this sphere. In other words, all three dimensions of space have been integrated into two dimensions, and these two dimensions exist on the surface of a sphere. The remaining dimension—time—is represented by the radius, or depth, of the sphere.

In this analogy, the Universe and all its contents should *not* be considered to be distributed inside the sphere. Rather, they are

FIGURE 26 A finite but unbounded sphere is one way to visualize a model of the entire Universe. All three dimensions of space are consolidated onto the (two-dimensional) surface of the sphere, while the fourth dimension, time, is represented by the radius of that sphere. On the *surface* of such a sphere, there is no boundary, edge, center, or special location; this is the basis of the cosmological principle, according to which all observers, everywhere in the Universe, are expected to observe pretty much the same thing.

spread *only on its surface*. All three dimensions of space are warped—in this special case, into a perfect sphere—because of the net influence of all the matter within every astronomical object. Thus, all the galaxies, stars, planets, and people, and even all the radiation, reside only on the surface of the sphere of this model Universe.

Now, since the radius of this model sphere represents time, we are forced to conclude that this spherical analogue grows with time. After all, as we discussed in chapter 1, the galaxies are observed to be receding. As time marches on, the radius of the sphere increases and so does its surface area, much like an inflating balloon. In this way, our three-dimensional analogue agrees with the observational fact that the Universe is expanding.

Actually, in 1917, Einstein did not know of the universal expansion. Hubble and other astronomers did not observationally establish the recession of the galaxies until the 1930s. Einstein's own field equations had allowed an expansion (or contraction) of the Universe, but he didn't believe it. Apparently, even Einstein was somewhat influenced by the (then) still-popular Aristotelian philosophy, which maintained that the cosmos (beyond the Moon's orbit) was immutable. So Einstein tinkered with his field equations, introduced a fudge factor that just offset the predicted expansion, and thus forced the model Universe to remain static. He was wrong in doing this; indeed, he later declared it the biggest blunder of his scientific career.

This error did not prevent Einstein and other relativists from uncovering many notable features of curved spacetime. One of the most intriguing findings is termed the cosmological principle —the notion that all observers perceive the Universe in roughly the same way regardless of their actual location. Put slightly more technically, the Universe is presumed to be homogeneous and isotropic.

To grasp the essence of the cosmological principle, consider a sphere again. It can be any sphere, so let it be Earth. Imagine

ourselves at some desolate location on Earth's surface, for instance, in the middle of the Pacific Ocean. To validate this analogy, we must confine ourselves to two dimensions of space; we can
look east or west, and north or south, but we cannot look up or
down. This is the life of a fictional "flatlander"—a person who
can visualize only two dimensions of space. Perceiving our surroundings, we note a very definite horizon everywhere. The surface *seems* flat and pretty much identical in all directions (even
though we know it is really curved). Accordingly, we might gain
the impression of being at the center of something. But we are
not really at the center of Earth's surface at all. *There is no center
on the surface of a sphere.* That is the crux of the cosmological
principle: there is no preferred, special, or central location on the
surface of a sphere.

Likewise, regardless of our position in the real, four-dimensional Universe, we observe roughly the same spread of galaxies
as would be noted by any other observer from any other vantage
point in the Universe. Despite our observation that galaxies literally surround us in the sky, this need not mean that we reside at
the center of the Universe. Indeed, if our spherical analogy is
valid, then there is no center in the Universe. Nor is there any
edge or boundary. A "flatlander" roaming forever on the surface
of a three-dimensional sphere seems completely analogous to a
space traveler (or any radiation) voyaging through the four-dimensional Universe. Neither ever reaches a boundary or an edge.
Proceeding far enough in a given direction on the surface of the
sphere, the traveler (or the radiation) would eventually return to
the starting point, just as Magellan's crew proved long ago by
circumnavigating the planet Earth.

In much the same way, if four-dimensional spacetime is structured according to this spherical analogy—and it might be—
then a beam of light or an astronaut can conceivably travel in one
direction, only to return at some future date from the opposite
direction. Imagine the freakish case of a (long-lived) observer

who, having launched a light beam in one direction, is eventually illuminated from the rear as the light returns from the opposite direction—about a hundred billion years later!

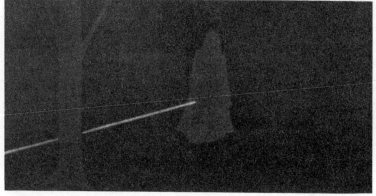

FIGURE 27 If the geometry of spacetime is (spherically) warped, as originally envisioned by Einstein, then it is possible (at least in principle) that a beam of light, launched in one direction, can someday return from the opposite direction.

Today, we realize that the Universe is not at all static. The recessional motions of the galaxies make indisputable the fact that it is expanding. Guided by the 1920s efforts of the Russian meteorologist Alexander Friedmann and the Belgian priest Georges Lemaître (who unsurprisingly were not collaborators), modern relativists seek more realistic models of the Universe by coupling Einstein's field and geodesic equations together with the observed rate of universal expansion. In this way, observations of galaxy recession become a boundary condition, a factual constraint helping us refine our latest cosmological models.

Note that the cosmological principle is valid even though the Universe is expanding; no surface of any expanding sphere, like that of any static sphere, has a center, edge, or boundary. To see this, imagine a sphere again, though now one that can swell. For example, visualize the entire Earth to be expanding, causing the surface area of our planet to increase as time progresses. Standing on such a hypothetically expanding "Earth," we would see familiar objects moving away; all surface objects—whether trees, homes, or mountains—would appear to recede. Now, more than ever, we might want to conclude that our position is special— that we exist at the center of some explosion. But we do not. Our position is no more special than anyone else's on the sphere's surface. In fact, everyone everywhere on the expanding surface would observe his or her surroundings to be receding. Who is correct? Everyone is correct. Recessional motions are observed from *any and all* positions on the surface of an expanding sphere.

Figure 28 further illustrates this all-important cosmological principle. Each frame shows a group of spirals painted onto the surface of a balloon. The spirals are meant to represent galaxies, whereas the balloon represents the Universe. Together the frames mimic a movie, a hypothetical filmstrip of numerous galaxies at successive times in the history of the Universe. Concentrate on the three spirals in the left-hand frame. Imagine yourself as a resident of one of these spirals, and note your position relative to nearby spirals. As the balloon inflates and the frames are progressively followed from left to right, the spirals seem to move away from each other as the Universe expands.

FIGURE 28 Galaxies in an expanding Universe appear to recede from one another regardless of the galaxy inhabited. This can be illustrated, at least in terms of our spherical analogue of Figure 26, by inflating a balloon; spots, drawn here as spirals on the surface of the balloon, recede from one another as the balloon inflates. Every observer in any galaxy would perceive all the other galaxies to be drifting away—and at roughly the same rate too. Thus, the cosmological principle holds valid even for a dynamically changing Universe. (However, unlike the painted spirals on the balloon, real galaxies have never been observed to be growing larger with time; this, among several other reasons, is why this spherical model of the Universe is only an analogy, not an example.)

The important point to realize here is that, regardless of which galaxy we inhabit, we would see roughly the same sort of galaxy distribution *and* the same sort of galaxy recession. To appreciate this, focus on a different spiral. Again track the positions of the other spirals while viewing the frames from left to right. The galaxies appear to recede for any observer in the Universe. Nothing is special or peculiar about the fact that we measure all the galaxies to be receding from *us*. This is so for *all* observers everywhere. As the cosmological principle again tells us, no observer anywhere in the Universe has a privileged position.

And so it is in the real, four-dimensional Universe. Although galaxies recede from the planet Earth, this is not a peculiarity of our vantage point: all observers anywhere in the Universe would see essentially the same sort of galaxy recession. Neither we nor any other beings reside at the center of the expanding Universe. Indeed, there is no real center in *space*. There is no position that we can ever hope to identify as the location from which the universal expansion began.

There is nonetheless a center in *time*. This is the origin of

time, and it corresponds in our three-dimensional spherical analogue to the sphere having zero radius. In other words, at the
beginning of the Universe, the three-dimensional sphere was a
point. It had a radius of zero. This was the origin of time. We
can think of it as the edge of time. But there is no edge in space.

Finally, recognize that no one really knows if the cosmological
principle is absolutely correct. Astronomers have adopted it as a
working hypothesis, largely because it vastly simplifies Einstein's
field equations. We can say only that the cosmological principle
seems consistent with all observations made thus far.

Visualizing the past when the sphere was much smaller, we
can ask an obvious question: When did the sphere have zero
radius—when was it a mere point? In other words, how long ago
were all the contents of the Universe squashed into a single speck?
More fundamentally, when did time begin?

To appreciate answers to these questions, imagine that time
can be reversed. Mentally reverse the expansion of the Universe
by contracting it at the same rate as we currently observe it to be
expanding. The galaxies would come together, eventually touch,
and finally mix. If we can estimate how long it would take for the
whole Universe to shrink to a single point, we shall then have a
measure of its age.

Actually, we can quantify this problem in very simple terms.
Since the distance traveled by any object equals the product of
its velocity and the time expended, Hubble's relation (see chapter
1) can be reexpressed as

$$\text{velocity} = \text{Hubble's constant} \times \text{velocity} \times \text{time}$$

Canceling the velocity terms from each side of the equation, we
find that

$$\text{time} = \frac{1}{\text{Hubble's constant}}$$

The inverse of Hubble's constant is thus a measure of the total time interval during which the cosmic objects have been receding from one another. It is the time since the expelled debris of universal matter reached the places at which they are now observed. This amount of time is then the answer to the fundamental question, How long ago in the distant past was all the matter

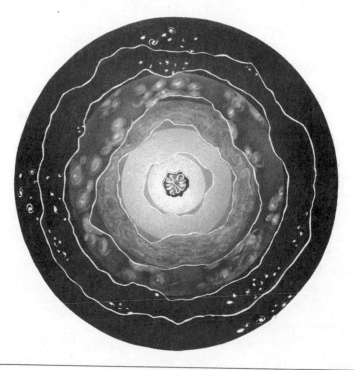

FIGURE 29 By combining the essential features of figures 26 and 28, we can depict a more accurate representation of the spherical analogue often used to model the Universe. As is shown in this cutaway illustration, the expanding Universe can be visualized better as a concentric series of spheres, each keyed to a different epoch of time. The innermost sphere, closest in time to creation, is the brightest, the outer spheres having become progressively darker (save for the galaxies of stars) as the cosmos cooled and thinned. The outermost sphere represents today, from our perspective on Earth, but we cannot look along this "current" sphere; "looking out is looking back," and as we scan the distant cosmos, we are thus examining objects as they once existed on earlier spheres farther inside Einstein's curveball.

in the Universe concentrated within a singular region of space? It is a measure of the duration of the Universe's expansion—quite simply, its age.

Numerically, the inverse of Hubble's constant yields a value of approximately fifteen billion years. Thus, the singular, compact region of space often associated with the origin of the Universe must have "exploded"—for whatever reason—about fifteen billion years ago. (Actually, as is explained a few chapters hence, the Universe might be a bit older, since its expansion is known to be slowing down.)

A final caveat: the range of possible error in this age is considerable since Hubble's constant is not known precisely. As we mentioned in chapter 1, astronomers now suspect that its value is somewhat affected by the "drift" of the local cluster of galaxies of which our Milky Way is a member—a motion that is currently uncertain and especially hard to measure accurately. In fact, the rate of galaxy recession depends critically on the rather uncertain distances of the farthest galaxies, which, owing to their faintness, are the most difficult ones to observe well. (The velocities of galaxies, near and far, are by contrast relatively easy to measure.) Currently, some researchers argue that the Universe could be as young as ten billion years, whereas others maintain that it is nearly as old as twenty billion years. An error of several billion years may seem large, but the difference between these extremes is only a factor of two, really quite good for an order-of-magnitude subject like cosmology. Compromising, we adopt fifteen billion years as the approximate age of the Universe—a remarkable finding in and of itself.

A Geometry for Spacetime

ELEVEN

Evolutionary Models of the Universe

I N THIS CHAPTER, we discuss how the Einsteinian field equations can be used to derive models of the entire Universe. These equations are central in deducing its bulk structure. Is the Universe flat and infinite, or negatively curved like the surface of a saddle, or perhaps positively curved like a surface of a sphere? These are among the possibilities for our global view of the Universe.

Figure 30 audaciously plots the size of the Universe against cosmic time, thereby making graphic the widest possible temporal perspective. By "size of the Universe," we mean either (in principle) the total four-dimensional region of spacetime in which the galaxies reside or (in practice) the average distance separating the galaxy clusters. Both notions validly represent the dimensions of the Universe on the grandest scale, but only the latter can be observationally measured.

The curve drawn in figure 30 shows two apparent facts: the

Universe began with an explosion of some sort, commonly called the big bang, after which its size increased with time. As we noted earlier, the Universe expands at a rate inversely proportional to the density of matter contained in it. After all, each clump of matter in the Universe gravitationally pulls on all other clumps of matter. Since this gravitational force is always attractive, it tends to counteract the expansion. So a tightly packed Universe —namely, one that is reasonably dense—causes a strong net gravitational pull, and a potentially significant slowing of the universal expansion. Even a rather tenuous Universe houses huge quantities of matter and would therefore be expected eventually to slow the rate of expansion, hence curve the solid line as drawn in figure 30. (Notice that I have returned to the notion of gravity; though warped spacetime is a better concept, I shall use the more familiar term, "gravity," whenever it makes the argument easier to comprehend.)

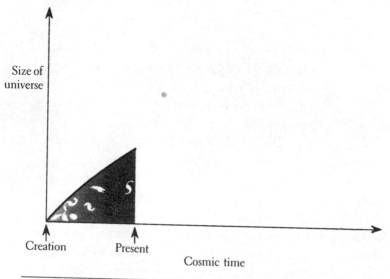

FIGURE 30 In this simple yet bold graph of universal size as a function of cosmic time, the darkened area represents the expansion of the Universe from its origin to the present.

The phenomenon of universal expansion uncannily resembles the three rocket cases considered in chapter 8. Each rocket recedes from its parent planet at a rate inversely proportional to the planet's mass. Earth, for example, pulled strongly on the launched rocket and was able to slow the rocket's escape; a less massive planet, Mars, slowed the rocket proportionately less; the most massive planet, Jupiter, exerted the strongest pull on the rocket and was able to halt its escape. The parallel between the orbital dynamics of a small rocket and the cosmic dynamics of the entire Universe is quite a good one. As for rockets, for a dynamic, changing Universe there are three conceivable models. Let us consider each of them in turn.

The first model Universe is one that evolves from a powerful initial "bang"—an explosion of some sort at the origin of time—and that thereafter expands forevermore. As the Universe grows from what must have been an extremely dense primeval clump of energy, the primordial stuff becomes progressively more diluted throughout the Universe, causing the average density of matter to decline. In this first model, insufficient matter exists to counteract the expansion, and the resulting Universe is analogous to the rocket moving away from Mars in figure 25. The total amount of matter cannot halt the outward motion of either the rocket or the Universe. Since this model Universe will theoretically arrive at infinity with some finite (nonzero) velocity, some researchers refer to this case as the hyperbolic model of the Universe. Its spacetime is curved like the surface of a saddle.

A hyperbolic model is said to imply an "open Universe." It is open in the sense that the initial bang was large enough, and the contained matter spread thinly enough, to ensure that this type of Universe will never stop expanding. Although matter everywhere pulls on all other parts of the Universe, this model Universe will never collapse back on itself. There is simply not enough matter.

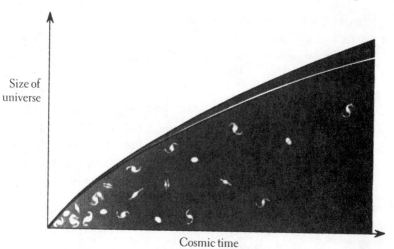

FIGURE 31 Both hyperbolic and parabolic galaxy motions imply an open Universe. The shaded area of this figure depicts how such an open Universe expands forevermore, the matter eternally thinning with the march of time. (The technically inclined reader will note that the hyperbolic model is represented by the entire darkened area of this figure, the parabolic model only by the area below the thin, white arc near the top of the diagram; both models will never stop expanding before reaching infinity.)

Of course, the Universe can never really become infinitely large. An infinite amount of time would be required to reach infinity. This is just the mathematician's way of stating that the hyperbolic, or open, Universe will expand endlessly. Properly stated, an open Universe *approaches* infinity.

If this model is correct, the galaxies will recede forevermore. With time, for an observer on Earth, they will fade toward invisibility, their radiation becoming weaker with increasing distance. Eventually, perhaps, even some of the closest galaxies will be so far away as to be hardly visible. Someday, all the galaxies might become unobservable; they will be too old and too distant, their radiation too faint. The Milky Way Galaxy will then be the only matter within the observable Universe. All else, even through the most powerful telescopes, will be dark and quiet. And even beyond that in time, the Milky Way will also someday peter out, as the hydrogen in all its stars changes into heavy elements. This type of Universe, including all its contents, eventually experiences a "cold death." The radiation, matter, and life in such a Universe are all destined to freeze.

Another open model of the Universe is conceivable, this one mimicking the parabolic trajectory of the Earth-launched rocket depicted in figure 25. In this case, the accumulated cosmic matter is precisely sufficient to halt the expansion eventually—but only after reaching infinity. This type of Universe also expands forever, though it barely makes it to infinity before exhausting its kinetic energy. But since nothing can ever really reach infinity, this parabolic model, like the hyperbolic model, represents an open Universe that will expand forever. Interestingly enough, its *net* spacetime is uncurved, or flat; Euclid would have loved it.

Yet another model for the Universe is plausible. Like the open ones just considered, this model also expands with time from a superdense original point. Unlike the open Universe, though, this Universe contains enough matter to halt the cosmic expansion before reaching infinity. That is to say, the outwardly expelled matter has been losing momentum ever since the initial bang, so much so that the galaxies will skid to a stop sometime in the future. Astronomers everywhere—on any planet within any galaxy—will then announce that the galaxies' radiation is no longer red shifted. The cosmological principle guarantees that this new view will prevail everywhere. The bulk motion of the Universe, and of the galaxies within, will be stilled, at least momentarily.

The expansion may well stop, but the inward pull of gravity does not. Gravitational attraction is relentless. Accordingly, this type of Universe will necessarily contract. It cannot stay motionless; nothing fails to change. Astronomers everywhere will then announce that the galaxies have blue shifts. (Actually, the change in perception will be gradual since different distances correspond to different times; accordingly, observers would probably notice that nearby galaxies are blue shifted even though distant ones are still red shifted.) In broadest terms, the contraction of this type of Universe is a mirror image of its expansion. Not an instantaneous collapse, it is rather a steady movement toward an ultimate end, requiring just as much time to fall back as it took to rise.

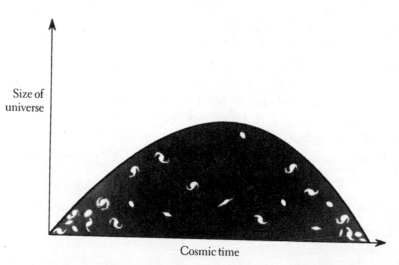

FIGURE 32 An elliptical, or closed, model of the Universe has a beginning, an end, and a finite lifetime. Matter, thinning from the effects of a big bang at the start, eventually returns to a point much like that from which the cosmos began, causing a "big crunch" at the end of all time.

This model in many ways resembles our earlier Jupiter analogy, for which the gravitational pull was great enough to cause the rocket's path to become elliptical. Having a similar geometrical pattern, a model of the Universe containing enough matter to reverse the expansion is often called an elliptical Universe. We also sometimes refer to it as a closed Universe—closed because it represents a Universe finite in size and in time. It has a beginning and an end. Its net spacetime mimics that of a sphere, as was described in the preceding chapter.

The expansion-contraction scenario of a closed Universe has many fascinating (and dire) implications. From what must have been a truly huge value initially, density drops to a rather small value by the time the Universe begins contracting, then returns to an enormously large value when, at some future epoch, all the matter collapses onto itself. Life, which has evolved steadily from simplicity to complexity during the expansion, will begin breaking down into simplicity again while inevitably heading toward its demise during the contraction. Why? Because toward the end of the contraction phase, the galaxies are destined to collide frequently as the total amount of space in which they exist diminishes. And just as compressing the air in a bicycle pump or rubbing our hands causes heating by way of friction, collisions among galaxies will generate heat as well. The entire Universe will grow progressively denser and hotter as the contraction approaches the end. Near the total collapse, the temperature of the entire Universe will have become greater than that of a typical star. Everything everywhere will have become bright—so bright, in fact, that stars themselves will cease to shine for want of contrasting darkness. This type of Universe will then shrink toward the superdense, superhot state of matter similar, if not identical, to the one from which it originated. In contrast to the open Universe that ends as a frozen cinder, this closed Universe will experience a "heat death." Its contents are destined to fry.

Cosmologists are uncertain of the fate of a closed Universe when it reaches this superhot, superdense, infinitely small state, known among scientists as a singularity. The Universe might simply end. Or it might bounce—into another cycle of expansion and contraction. Frankly, the mathematics of singularities has not yet been fathomed. This ultimate state of matter poses one of the hardest problems in all of science. For the most part, physicists and astronomers are experimentally and theoretically ignorant of the physics of singularities. Here's the problem.

As both density and temperature increase while the contraction nears completion, pressure—the product of density and temperature, at least in classical physics—must increase phenomenally. The question as yet unanswered is, Will the Universe just end as a final, minuscule speck, or will this pressure be sufficient to overwhelm the relentless pull of gravity, thereby pushing the Universe back out into another cycle of expansion and contraction? In other words, will a closed Universe rebound? If so, it may well have many—perhaps an infinite number of—periods of expansion and contraction. Often termed an oscillating Universe, it is in some ways a hybrid of the open and closed models.

Unquestionably, a certain amount of aesthetic beauty embel-

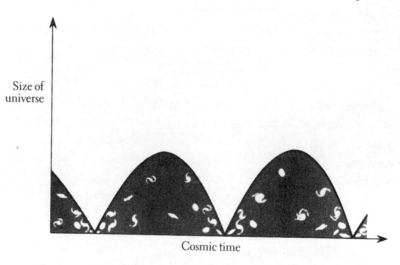

FIGURE 33 An oscillating Universe compromises the basic features of the open (indefinitely long) and closed (beginning and end) models. Here, matter alternatively thins and cools, and compresses and heats, again and again. Such a Universe experiences small bangs from time to time, but it did not have a unique beginning, nor will it ever truly end. This kind of Universe always was and always will be—a philosophically satisfying notion for many people.

lishes this model of the Universe. Here, there is no need for a unique, once-and-for-all-time explosion—no need for a *big* bang. Nor does this model embody a definite beginning or a definite end. The oscillating model merely goes through phases, each initiated by a separate explosion, or bang. Indeed, an oscillating model has many "bangs," each expansion a "day," each contraction a "night." But none of these bangs is unique, none of the origins apparently any more significant than any other. Furthermore, the idea of oscillation avoids the potential philosophical problem of what preceded a unique big bang of a one-cycle closed Universe or of an open Universe.

If the oscillating model is valid, we need not trouble ourselves with the concept of "existence" before the beginning of time. Such a Universe always was and always will be.

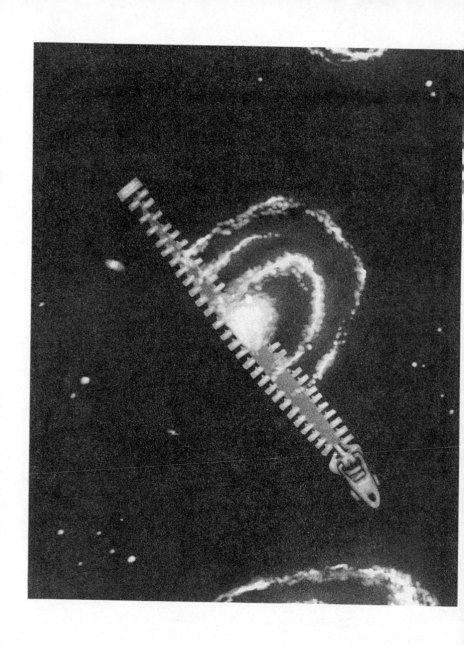

Out of the Blue

Other Universe Models

ALL THREE MODELS of the Universe discussed in the preceding chapter take evolution as their guiding principle: the Universe changes with time. They are derived from Einstein's general theory of relativity and are favored in one form or another by the great majority of today's cosmologists. However, several other models of the Universe have been proposed during the past few decades. Most of them do not follow directly from relativity; some do not even call for change with time or embrace evolution as their central theme. It is worth considering one of the more prominent ones, for until recently it was favored by some segments of the scientific community.

The "steady-state" model stipulates that the Universe appears roughly the same not only for all observers but also for all time. Its fundamental tenet is embodied in what is sometimes called the *perfect* cosmological principle: to any observer at any time, the physical state of the Universe is the same. In other words, the average density of the Universe remains eternally constant. It holds *steady*.

The steady-state model was conceived in (the real) Cambridge by Herman Bondi, Thomas Gold, and Fred Hoyle as a powerful alternative to the various time-dependent, or evolutionary, models of the Universe. Its initial motivation was based as much on philosophy as on science. The oscillating Universe aside, many scientists and philosophers were (and still are) unwilling to concede that nothing could have existed prior to a unique big bang. To ask what preceded the origin of the Universe is to make one of the trickiest queries imaginable. What existed before the big bang? Why was there a big bang? What or who caused it? These are questions unaddressable within the realm of contemporary science. When there are no data, the scientific method becomes a useless investigative technique. Philosophies, religions, and cults of all sorts can offer theories to the nth degree, but science remains mute.

The steady-state model avoids these thorny questions, as does the oscillating model. For them, there is neither a beginning nor an end. The Universe just *is* for all time.

Steady-state cosmologists concede that the Universe is expanding, for they are unable to refute the observational fact that the galaxies are receding. Finding anathema the concept of an expansion as the aftermath of an explosion, they are forced to postulate an unknown repulsive force to push the galaxies apart. Still, the principal objective of the steady-statists is that the bulk view of the Universe—the average density of matter and the average distance between galaxy clusters—remains constant forever. Accordingly, to offset the dilution of the density due to the galaxies' recession, the steady-state model requires the emergence of new matter in the Universe. Odd as it may seem, the steady-state theorists propose that this new matter is created from nothing. Even so, and despite the recession of the galaxies, the creation of additional galaxies in just the right amount can keep constant the number of galaxies per unit volume. In this way, the average density of matter in the Universe is preserved forever.

The most vexing problem with the steady-state model is its failure to specify how the additional matter is created. Nor does it specify where. Some researchers theorize its infusion out beyond the galaxies in intergalactic space, whereas others prefer injection in the centers of galaxies. Not much new matter is needed to offset the natural thinning of matter as the galaxies speed apart. The creation of a single hydrogen atom in a volume roughly equivalent to that of the Houston Astrodome every few years would suffice. Unfortunately, the sudden appearance of such a minute quantity of matter, either inside or outside of galaxies, is currently quite impossible to detect and, therefore, to test.

Regardless of *where* matter is created, the real problem is *how* it's created. The sudden appearance of new matter from nothing violates one of the most cherished concepts of modern physics — the conservation of matter and energy. A widely embraced dogma of contemporary science maintains that the sum of all matter and all energy is constant throughout the Universe. Matter can indeed be created from energy (and energy from matter), but it is very

FIGURE 34 Because of the recession of the galaxies, the steady-state model of the Universe requires the continual creation of new matter in order to keep the bulk appearance of the Universe constant for all time. In these three frames, time drifts from left to right; as the galaxies separate, new matter somehow somewhere comes into being, causing the density of galaxies to remain roughly the same in all three frames. This kind of Universe is very much out of favor among contemporary thinkers.

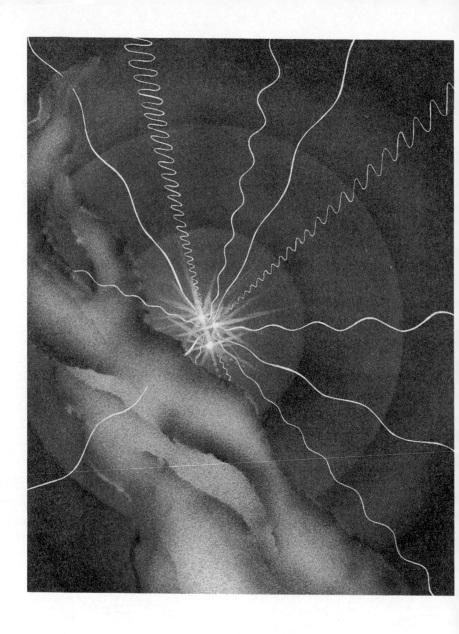

DEGREES OF RADIATION

THIRTEEN

Cosmological Tests

HOW CAN WE distinguish among the various conceivable models of the Universe discussed in the two preceding chapters? The most straightforward way uses the process of elimination.

The steady-state model seems untenable for at least two reasons. First, the spread of galaxies is not uniform throughout space. Provided the red shift is a true indicator of distance, the faraway active galaxies (called quasars) far outnumber (per unit volume) those nearby. Had we lived ten billion years ago, when quasars were presumably the dominant astronomical objects, our view would have been filled with quasars—many more than now surround our vantage point on Earth. The perfect cosmological principle seems clearly violated: the large-scale view of the Universe was not the same eons ago as it is now.

A second, rather fatal argument against the steady-state model is the serendipitous result of experimentation. Observations made with radio telescopes always yield a signal, regardless of the time of day or night. Unlike optical observations, in which we sometimes encounter a complete void of light toward the dark, obscured regions of space, radio observations never fail to detect some radiation. Sometimes the radio signal is strong, especially when the telescope is aimed toward an obvious source of radio emission. At other times it is weak, particularly in regions devoid of all known radio sources. Yet, whenever the accumulated emission from all the known radio sources and from all the atmospheric and instrumental noise is accounted for, there remains a minute radio signal—a sort of a weak hiss much like static on a home AM radio or the "snow" seen on an inactive channel of a television set. Never diminishing or intensifying, this weak signal is detectable at any time of the day, on any day of the year, year after year; it's omnipresent, apparently inundating all of space. Moreover, it is equally intense in any direction of the sky—that is, isotropic. The whole Universe is seemingly awash in this feeble radio radiation.

This ubiquitous signal was accidentally discovered with a radio antenna that two Bell Laboratories scientists, Arno Penzias and Robert Wilson, were using in the mid-1960s while attempting to map certain low-frequency galactic radio emissions. In their data, they unexpectedly noticed a bothersome radio hiss that just would not go away. Unaware that they had detected a signal of cosmological significance, the researchers sought many different origins for the excess emission, including atmospheric storms, ground interference, equipment short circuits, even pigeon droppings deposited inside the horn-shaped antenna structure. Later conversations with theorists at Princeton University enlightened the experimentalists about the static's most probable origin. That origin is the fiery creation of the Universe itself.

This weak, isotropic radio radiation is widely interpreted as a veritable "fossil" of the primeval explosion that began the universal expansion long ago. The relic hiss is often termed the cosmic background radiation. Its existence is completely consistent with

FIGURE 35 This "sugarscoop" antenna in New Jersey, originally built by Bell Labs technicians to communicate with unmanned satellites, was used to detect and study the cosmic background radiation that harks back to the creation of the Universe. The device acts as a big ear, collecting radio waves in the aperture at the right, guiding the radiation to a focus at the left, and detecting the received signals within the blockhouse for further analysis. The entire apparatus rotates on a circular railroad track.

any of the evolutionary models of the Universe, but there is no clear role for it in the steady-state model.

The cosmic background radiation is presumed to be a remnant of an extremely hot phase of the early Universe—a Universe that has cooled during the past fifteen billion years or so. Regardless of whether the initial explosion was a unique big bang producing an open and infinite Universe or a closed and finite one, or even one of repeated bangs characterizing an oscillating Universe, the primeval, superhot, superdense matter must have emitted high-frequency thermal radiation. All objects having any heat release energy; a very hot piece of metal (a branding iron, for instance) glows with a red- or white-hot brilliance, whereas less-hot metal (such as a laundry iron) feels warm to the touch and emits less-energetic infrared or radio radiation. In its fiery beginnings, the Universe almost certainly released extremely energetic gamma-ray radiation. But with time the Universe expanded, thinned, and cooled. Consequently, the emitted radiation must have steadily shifted from the gamma- and X-ray varieties normally associated with intensely hot matter, down through the less-energetic ultraviolet, visible, and infrared kinds, eventually becoming the radio waves usually released by relatively cool matter.

The evolutionary models predict that some fifteen billion years after the primeval explosion, the average temperature of the Universe—the relic of this big bang—should be quite low, in fact no more than about −270 degrees Celsius (or about a few degrees Kelvin). That is far below zero degrees Celsius, the temperature at which water freezes, and only a few degrees above the absolutely lowest value (zero degrees Kelvin), at which all atomic and molecular motions virtually cease.

To confirm the theory, researchers have carefully measured the intensity of the weak isotropic radio signal at a variety of frequencies. Figure 37 plots the results of these observations. The dashed line is the best-fit solution for all the data acquired during the past decade; confirming the theory, the curve shows a universal temperature of approximately −270 degrees Celsius (or 3 degrees Kelvin). Furthermore, this oldest fossil really does seem to pervade the whole Universe, including Earth and the building and the room where you are now reading this. The amount of cosmic radiation absorbed by humans at any one time, however, is minuscule, totaling less than a billionth of the power emitted by a hundred-watt light bulb.

So, the existence of the cosmic background radiation and the spread of galaxies in space discredit the steady-state hypothesis as a feasible model of the Universe. Clearly, the Universe has

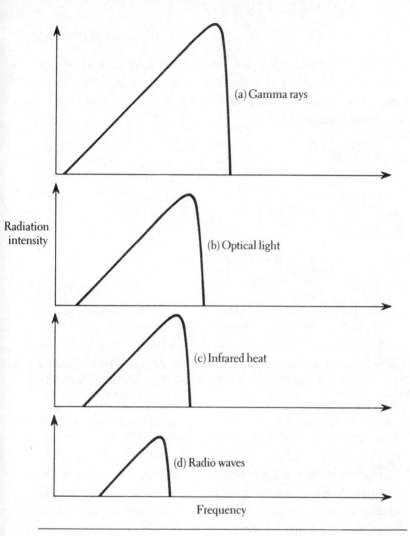

Radiation intensity

(a) Gamma rays

(b) Optical light

(c) Infrared heat

(d) Radio waves

Frequency

FIGURE 36 These theoretically derived curves show how the intensity of radiation varies as a function of frequency. They characterize the total emission of radiation from the entire Universe (a) shortly after the origin of the Universe, when it was flooded with X and gamma rays, (b) some five billion and (c) ten billion years thereafter, when the radiation had been downgraded to optical and infrared waves, and (d) at present, approximately fifteen billion years after the primeval explosion. Though we can think of today's cosmic background radio radiation as being a cooled relic of the creation fireball, we are more correct to regard it as the greatly red-shifted glare of the intensely energetic conditions prevalent at the start of the Universe.

changed with time; it has not been steady at all. Our choice of the correct model of the Universe must then be made from among the various evolutionary models. We need other data to sift through each of them.

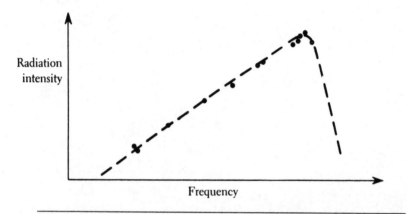

FIGURE 37 Observations *(filled dots)* of the cosmic background radiation agree well with those expected from theory (consult frame *d* of figure 36). The dashed curve is the best fit to the data, consistent with a thermal temperature of only a few degrees above absolute zero. The raw data plotted in this figure represent some of the most profound information ever acquired by humans; these data take us back closer to creation than any other facts extant in recorded history.

The most straightforward way to distinguish between the open and closed models requires an estimate of the average density of matter in the Universe. More than anything else, density is what differentiates the closed model, wherein matter is packed tightly enough to halt gravitationally (and mutually) the universal expansion before it reaches infinity, from the open model, wherein the density is simply not large enough to bring the Universe back.

With relativity in mind, we can compute the precise density of matter needed to halt the expansion just as the outer limits of the Universe reach infinity. For today's thinned-out Universe, the answer is roughly 10^{-30} gram/cubic centimeter. (Water, by contrast, has a density of 1 gram/cubic centimeter.) This extraordinarily small density equals hardly more than 10^{-6} hydrogen atom/cubic centimeter or, equivalently, a few hydrogen atoms within a volume the size of a typical household closet. Terribly tenuous, this is in fact many orders of magnitude thinner than the best vacuum attainable in terrestrial physics laboratories. But remember, this is an *average* density of the *entire* Universe— calculated by lumping the galaxy clusters, where the matter is most concentrated, with the intergalactic voids, where little if any matter resides. In short, it is the critical universal density corresponding to the parabolic model, which is, so to speak, intermediate to the hyperbolic and elliptical models of the Universe.

We can then ask, What is the actual density of matter in the Universe? Is it more or less than this critical value? Cosmologists quantify these questions by letting the Greek symbol Ω (omega) denote the following ratio:

$$\Omega = \frac{\text{actual density}}{10^{-30}}$$

If Ω equals 1 precisely, then the actual density equals the above theoretically computed density, and the Universe obeys the parabolic model; it is open and will continue to expand forever (though theoretically it will reach infinity and then stop expanding). This model dictates that the Universe has no net curvature, that it is a flat Universe governed largely by Euclidean geometry and Newtonian dynamics. Localized regions of spacetime, especially those near massive astronomical objects, are surely curved, but on the whole the accumulated curvature of spacetime is zero.

If the value of Ω is less than 1, then the Universe's matter is not dense enough ever to stop the expansion (even at infinity). This type of Universe is destined to expand forever, thus conforming to the open, infinite, hyperbolic model. If, on the other hand, Ω exceeds 1, then the closed, finite, elliptical Universe prevails, and it will someday start contracting. In either case—Ω greater or less than 1—Einstein's relativity rules.

Theory aside, how can we determine the value of Ω? At first, it seems simple. Just measure the average mass of each of the galaxies residing within any parcel of space (such as that of figure 2), estimate the volume of that space, and calculate the total density. Having done this many times, astronomers usually find about 10^{-31} gram/cubic centimeter. As best as can be determined, this computation is independent of whether the chosen region contains only a few galaxies or a rich cluster of them; the resulting density is pretty much the same, within a factor of two or three.

Galaxy-counting exercises thus favor $\Omega \simeq 0.1$, implying that the Universe is open. If this is correct, the Universe must have originated from a unique big bang and will expand forever. Such a Universe has no end but definitely had a beginning.

However, an important caveat deserves mention here. Observations during the past few years have demonstrated that all the matter in the Universe is not housed exclusively within brightly visible galaxies. Much dark matter seems to lurk beyond each of the galaxies. This is especially true for the invisible regions within the galaxy clusters like that of figure 2, where X-ray observations have proved the existence of heretofore unrecognized (hot but thin) chaotic gas. Furthermore, recent optical and radio observations are now hinting at the presence of huge invisible (cold but thin) halos of loose gas surrounding most galaxies. (To be sure, earlier theoretical studies had strongly implied the existence of additional dark matter; without it—indeed, without even more of it not yet found—the galaxy clusters would probably disperse, thus contradicting the observational inference that they are mutually bound aggregates.)

Our above estimate for Ω did not account for any of this "hidden" matter, since the extent and amount of it is currently unclear. But if more than ten times as much additional matter resides outside the galaxies as within the galaxies themselves, Ω will correspondingly increase by more than a factor of ten. That's why one of the most significant issues in frontier science today is to determine the amount of matter dwelling amid the dark realms inside galaxy clusters. For if substantive reservoirs of dark matter do skirt the galaxies, Ω could exceed 1, forecasting that the Universe is closed, having an end as well as a beginning. Whether such a Universe originated in a unique big bang, before which nothing existed, and whether such a Universe ends for all time without bouncing cannot be addressed by taking this kind of an inventory.

The value of Ω obtained by this galaxy-counting method is thus quite uncertain at present. It cannot unambiguously distinguish between the open and closed models, though at face value it favors an open Universe destined to expand forevermore.

Cosmologists have developed another observational test to determine the value of Ω. Like the method above, this test attempts to estimate the average density of the Universe. It essentially relies on the fact that each and every piece of matter gravitationally pulls on all other pieces of matter. Specifically, this second test addresses the question, At what rate is matter everywhere causing the universal expansion to slow down? Put another way, how fast is the Universe decelerating?

Because this cosmic slowdown is likely to be excruciatingly subtle, astronomers cannot realistically expect to measure it by watching the slackening motion of any one galaxy. But, by thinking more broadly, they have developed a statistical way to do it. If the Universe began in an explosive bang, it must have expanded rapidly at first and gradually grown more sluggish thereafter. The expansion of anything—an ordinary bomb, an atmospheric thunderclap, whatever—is always greater at the moment of explosion than at some later time. Hence, given that looking out into space is equivalent to probing back into time, the recession of the galaxies should be larger for the faraway galaxies and somewhat smaller for those nearby.

This argument suggests that we might be able to look back sufficiently far into the past to measure a greater expansion rate at an earlier epoch. Figure 38 illustrates how this greater expansion rate of earlier times would be expected to show departures from the usual Hubble relation. For nearby objects—namely, for galaxies whose radiation was emitted in relatively recent times— the usual Hubble relation (solid line) yields a certain amount of recessional velocity per distance interval. As we noted in chapter 1, our best estimate for the value of this line is Hubble's constant, and it numerically equals 22 kilometers/second for every million light-years of distance. If the universal expansion rate never changes, then Hubble's constant is a true constant for all time, and the solid line of figure 38 can be linearly extended to arbitrarily large distances; the (steady) solid line then corresponds to the now-discredited steady-state model of the Universe. By contrast, all the time-dependent or evolutionary models demand a greater expansion rate in the past—which means that Hubble's constant is not really constant. It must have been larger in earlier epochs of the Universe. How much larger depends upon the average density of matter in the Universe.

So, all the evolutionary models predict departures from Hubble's relation over the course of time. Drawn as dashed curves in figure 38, this departure is expected to be largest for the finite, closed model of the Universe, $\Omega > 1$; this is reasonable since the huge amounts of matter needed to close the Universe would significantly decelerate the cosmic expansion throughout the history of the Universe. The infinite, open models, $\Omega \leq 1$, are expected to show smaller departures, although they, too, predict more rapid expansion at earlier epochs. (Cosmic deceleration is the reason that *all* the evolutionary models display some slowing, as is exemplified by the gentle curving of the graph in figure 30.)

What do the data indicate? Is there any evidence for higher-than-Hubble recessional velocities among the more distant galaxies? The Hubble diagrams that plot the observed velocities of hundreds of galaxies having reasonably well determined distances

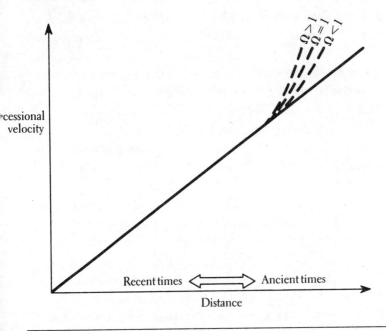

FIGURE 38 This Hubble diagram illustrates how each of the evolutionary models of the Universe is expected to show evidence for cosmic deceleration. The departures *(dashed curves)* from the usual Hubble relation *(solid line)* are exaggerated for clarity.

(out to a few billion light-years) demonstrate no clear departure. Nonetheless, astronomers have striven to extend those plots to greater distances, for galaxies do reside at remoter realms of the Universe. Admittedly, observations of such extraordinarily faint objects are difficult to make and, furthermore, do not take account of any (as yet undiscovered) evolutionary changes that might affect the brightness of galaxies over the course of time. Doubtless, the data are currently subject to considerable uncertainty, but most very distant galaxies do indeed have measurably greater recessional velocities than those expected from the Hubble relation. And although several leading astronomers seem to be vacillating on the verdict, these data prima facie obey the curve $\Omega > 1$, implying a closed and finite Universe. This second cosmological test then contradicts the first, unless substantial amounts of dark matter really do exist beyond the visible galaxies.

Astronomers have yet a third, albeit indirect, method of determining the fate of the Universe. This one depends sensitively on the number of nuclei synthesized shortly after creation. Though we are unsure how much helium was produced in the hot and dense primordial fireball of the early Universe, some must have been fashioned at that time, since there is just too much of it around to have been fused exclusively later in the cores of stars. Part of the problem lies in the uncertainty of the rate at which the Universe expanded during its earliest epochs. Put another way, we are unsure how fast the early Universe thinned and cooled. Time is of the essence here, for the slower the expansion, the greater the chance that elementary particles interacted with one another to fabricate helium. Most cosmologists suggest that there was probably enough time to create one helium nucleus for every fifteen hydrogen nuclei. This compares favorably to the observed value of about one for every twelve.

If there were no other way to produce helium, astronomers could conceivably use its observed abundance to turn the problem around and thereby infer the rate of universal expansion. In this way, today's cosmic abundance of helium might be able to tell us something about the early Universe. But the problem is not that simple, for helium nuclei are also produced in the hot and dense cores of stars. And it is virtually impossible to unravel the relative contributions to helium production made by primordial and stellar nucleosynthesis.

Deuterium, the heavy isotope of hydrogen (consisting of a proton and a neutron in nuclear form), is also created in the early Universe. Unlike helium, deuterium is not likely to be produced in stars. Therefore, our models of primordial nucleosynthesis should enable us to predict the amount of deuterium produced in the aftermath of the bang; these models suggest, on the average, one deuterium atom for each 100,000 hydrogen atoms.

Slightly more deuterium could have been synthesized if the Universe expanded just a bit faster than at a certain rate and slightly less deuterium if the Universe expanded just a little slower. In effect, the formation of deuterium depends sensitively upon the density of matter.

Observations of deuterium should then provide a test of the universal expansion rate. In turn, this should tell us if the Universe is open or closed, for if the rate is high the galaxies should recede forevermore, whereas if the rate is low the Universe might eventually contract. In recent years, some limited observations of deuterium have been accomplished, especially by means of orbiting satellites that can capture deuterium's strong spectral feature in the ultraviolet part of the electromagnetic spectrum. Though somewhat controversial, the most straightforward interpretation of the observations made of both interstellar and interplanetary gases implies that the Universe is open.

Whether we live in an open or closed Universe, then, is currently unknown. The bottom line clearly suggests an evolutionary Universe, but its ultimate destiny remains concealed. Most cosmologists are inclined to say that we should expect a definite answer within a few years. This is perhaps overly optimistic, for the final solution requires the agreement of three often disparate groups of human beings. First, there are the theoreticians, whose imaginative minds invent the model Universes; they try to determine what the Universe is supposed to be like. Second, there are the experimentalists, constantly testing the theories, all the while extending their observations to more distant realms in our varied Universe; they try to determine what the Universe really is like. And third, there are the skeptics, who regard the models of the first group as mere speculation and the results of the second group as overinterpretation of the data without due regard for observational error. In the end, all three attitudes are helpful and necessary, for only by their cooperation and counteraction can we ever hope to approach the truth.

BLACK HOLES

How are we to proceed from this point in order to obtain a complete theory of atomically constructed matter? In such a theory, singularities must certainly be excluded, since without such exclusion the differential equations do not completely determine the total field. Here, in the field theory of general relativity, we meet the same problem of a theoretical field-representation of matter as was met originally with the pure Maxwell theory.

Here again the attempt to construct particles out of the field theory, leads apparently to singularities. Here also the endeavor has been made to overcome this defect by the introduction of new field variables and by elaborating and extending the system of field equations. Recently, however, I discovered . . . that the above mentioned simplest combination of the field equations of gravitation and electricity produces centrally symmetrical solutions which can be represented as free of singularity (the well known centrally symmetrical solutions of Schwarzschild for the pure gravitational field, and those of Reissner for the electric field with consideration of its gravitational action). . . . In this way it seems possible to get for matter and its interactions a pure field theory free of additional hypotheses, one moreover whose test by submission to facts of experience does not result in difficulties other than purely mathematical ones, which difficulties, however, are very serious.

—A. Einstein, Out of My Later Years

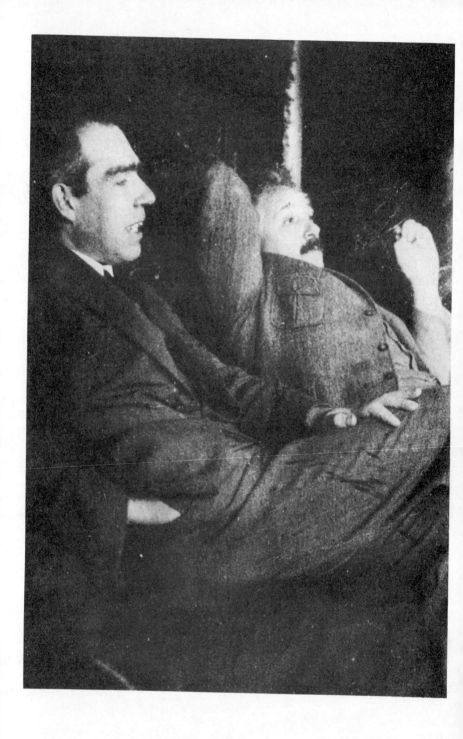

FOURTEEN

Nature of the Singularity

"LOOKING OUT into space is equivalent to looking back into time." That statement has been repeated many times now, and we must keep it in mind throughout this book. Just how far back in time can we probe? Is there any way to study the Universe beyond the most distant galaxy? How closely can we perceive the edge of time, the very act of creation itself?

Somewhere beyond the realm of the farthest galaxy (and thus long ago in time) lies a primeval plasma from which those galaxies evolved. This is not a structured object of any kind—just a hot, dense gas uniformly filling the entire Universe. It is the "stuff" that existed prior to the formation of galaxies, stars, planets, or any other organized cosmic object. Can we observe radiation from

Einstein with Niels Bohr

this most ancient gas? The answer is yes; astronomers have already done so.

The cosmic background radiation, detected by radio telescopes and discussed in the preceding chapter, is the "fossilized" remnant of the primeval plasma—the oldest known relic, we might say. Though it now engulfs us, this radiation originated far, far away in time—about 14.9995 billion years ago, or some 500,-000 years after creation. Hence, the radiation we detect is greatly red shifted or stretched in wavelength. In fact, radiation now reaching us from the exceedingly hot primeval plasma has been shifted clear across the electromagnetic spectrum. Intense gamma-ray radiation present soon after the birth of the Universe has been "downgraded" to rather feeble radio radiation during the past 15 billion years or so; the characteristic radiative pattern for a gas temperature of more than a billion degrees Celsius has been red shifted to the point where we now observe it to be about −270 degrees Celsius. In short, the lethal high-energy photons that escaped the gas of the original "bang" have found their way to us as harmless radio waves, just as the model of an expanding Universe that is both thinning and cooling leads us to expect.

So, we can imagine the cosmic background radiation as having been emitted by a cooled and darkened version of the primeval gas, as was explained in chapter 13. But we are more correct to regard this isotropic radio hiss as the red-shifted glare of the intensely hot conditions prevalent shortly after the start of the Universe. We can then better appreciate that the cosmic background radiation comes from the most remote region of the Universe that astronomers have thus far been able to probe observationally—remote in space and remote in time. Einstein would have been proud.

To appreciate even earlier epochs of the Universe, we must think deeply about times long, long ago. We must strive to imagine what it was like long before Earth, the Sun, or any of the stars materialized, even before the cosmic background radiation was emitted in the aftermath of the big bang. Some people have trouble imagining such truly ancient times. Fortunately, a small trick can help us comprehend the earliest moments of the Universe.

The "trick" involves the natural symmetry of the closed model of the Universe. It is this: if we find troublesome the need mentally to reverse time in order to appreciate the earliest universal epochs that *have* occurred, we can instead visualize the events that *will* occur as a closed Universe nears the final phase of its collapse. The physical conditions near the end of a closed Universe are expected to be virtually identical to those near the start because the mathematics describing the contraction is a mirror image of that describing the expansion. In other words, the events that *will* occur just prior to the end of a closed Universe mimic those that *already* happened just after its beginning. By using the laws of physics to predict the final events of a closed Universe, we can gain some inkling of the early aftermath of the universal bang approximately fifteen billion years ago.

Recall from figure 32 that a closed Universe will expand ever more slowly into the future, gradually coming to a halt. Once its contraction begins, the mean separation between the galaxies will decrease, causing more frequent galaxy collisions, larger amounts of friction, and higher temperatures for the whole Universe. The present cosmic background radiation will steadily rise as the Universe gradually compresses, just as any gas heats when the number of atomic collisions increases. (Witness the heating when compressing the air molecules inside a bicycle pump.) Eventually, the matter in such a contracting Universe becomes very hot and dense, culminating in a blazing, compact singularity. Such a bizarre region, having all the mass in the Universe concentrated in the volume of a pinhead, might be unable to exist in actuality. Even so, that is the state predicted by the currently known laws of physics.

The important item to note here is the complete symmetry between the origin and the end of time for a closed type of Universe. The superhot, superdense physical conditions expected to characterize the final singularity must have also typified the initial singularity from which the Universe emerged some fifteen billion years ago.

Einstein would also have been proud to realize that his quest to unify all the known forces of nature has recently borne fruit; some aspects of macroscopic astronomy and microscopic physics have seemingly been synthesized, thus creating a whole new inter-disciplinary specialty, termed particle cosmology. In particular, the electromagnetic force binding atoms and molecules and the weak nuclear force governing the decay of radioactive matter have been merged by a theory that asserts them to be different manifes-tations of one and the same force—an "electroweak" force. Cru-cial parts of this theory have recently been confirmed by experi-menters using the world's most powerful accelerator, in Geneva (shown partly in figure 19), and concerted efforts are now under way to extend this unified theory to include the strong nuclear force that binds elementary particles within nuclei. Furthermore, though physicists are unsure at this time how in turn to incorpo-rate into this comprehensive theory the fourth known force (grav-ity), there is reason to suspect that we are nearing the realization of Einstein's dream—the understanding of all the forces of na-ture as different aspects of a single, fundamental force.

Here is a brief, capsulated explanation of the operation of the newly understood electroweak force. In microscopic (quantum) physics, forces between two elementary (subatomic) particles are represented by the exchange of another type of particle called a boson; in effect, the two particles can be imagined as playing a rapid game of catch using a boson as a ball. In ordinary electro-magnetism familiar to us in our terrestrial surroundings, the boson is the usual photon—a bundle of energy that always travels at the speed of light. The new electroweak theory includes four such bosons: the usual photon as well as three other particles having the innocuous designations of W^+, W^-, and Z°. At relatively low temperatures—roughly below a million billion degrees, which is the range encompassing virtually everything we know about on Earth and in the stars—these bosons split into two families: that of the photon, which expresses the usual electromagnetic force, and that of the other three, which carry the weak force. But at temperatures higher than 10^{15} degrees, these bosons work together in such a way as to make indistinguishable the weak and electromagnetic forces. Thus, by experimentally studying the behavior of this new force, we gain insight into not only the essence of nature's building blocks but also the early epochs of the Universe, when the temperature was indeed that high.

To appreciate the nature of matter at temperatures substantially higher than 10^{15} degrees, and thereby to explore indirectly times even closer to creation, physicists are now researching a more general theory that incorporates the electroweak and strong nuclear forces (but not yet gravity). Several versions of this "grand-unified theory," dubbed GUT for short, have been proposed, though experimentation capable of determining which if any of these theories is correct has really only begun. Like the other forces just discussed, this grand force is mediated by a boson elementary particle, in this case called the X boson. It is, according to these grand theories, the very massive (and thus energetic) X bosons that play a vital role in the first instants of time.

Imagine a time just 10^{-39} second after creation, when the temperature was some 10^{30} degrees. At that moment, only one type of force other than gravitation operated—the grand unified force just noted. According to the theory of such a force, the matter of the Universe must have exerted a very high pressure that pushed outward in all directions. (In classical physics, pressure is the product of density and temperature, so if, in the early Universe, each of these quantities was large, then the pressure must have been vast.) The Universe must have responded to this pressure by expanding in a regular way, as described in earlier sections of this book; the temperature dropped as the Universe ballooned, in a manner inversely proportional to the size of the Universe. Thus, for example, as time advanced from 10^{-39} second to 10^{-35} second, the Universe grew another couple of orders of magnitude and the temperature fell to 10^{28} degrees.

Now, according to most grand-unified theories, this temperature—10^{28} degrees—is special, for at this value a dramatic change occurs in the expansion of the Universe. Briefly, when matter is "cooler" than this temperature, the X bosons can no longer be produced; after 10^{-35} second, the energy needed to create such particles was too dispersed, owing to the diminishing temperature. As the temperature fell below 10^{28} degrees, the X bosons disappeared, and instead of exerting an outward pressure, the matter of the Universe suddenly began to develop a huge inward tension (or negative pressure). Intuitively, we might reason that such an enormous inward force would halt the expansion of the Universe or at least slow it down. But general relativity teaches us differently; the actual effect of the sudden demise of the X bosons was a surge of energy roughly like that released as latent heat when water freezes (an event that occasionally bursts a closed container in the process). After all, energy no longer concentrated enough to yield X bosons was nonetheless available to enhance the general expansion of the Universe—in fact, to cause it to expand violently or "burst" for a short duration just after the demise of the bosons. The youthful Universe, though incredibly hot, was quite definitely cooling and in this way experienced a series of such "freezings" while passing progressively toward cooler states of being. Perhaps the most impressive of all such transitions, the change from outward pressure to inward tension caused a rapid acceleration in the rate of expansion. This period of (actually exponential) expansion has been popularly termed inflation; in a mere 10^{-35} second, the Universe inflated some twenty orders of magnitude or more (that is, more than a billion billion times).

At the conclusion of the inflationary phase, some 10^{-35} second after creation, the X bosons had disappeared forever, and with them the grand-unified force. In its place were the electroweak and strong nuclear forces that operate around us in the more familiar, lower-temperature Universe of today. With these new forces in control (along with gravity), the Universe once again experienced an outward pressure and thus resumed its more "leisurely" expansion.

What about even earlier phases of the Universe, those before 10^{-35} second? Can we probe, even theoretically, any closer to that maelstrom called creation, which occurred—by definition—at a time of zero seconds, the celebrated "$t = 0$" moment? Efforts are currently hampered because physicists are uncertain how to incorporate the gravitational force into the correct GUT. To be sure, no one has yet succeeded in developing a "super-grand" unified theory (or "super-GUT" in frontier parlance), as this is tantamount to inventing a quantum theory of gravity. Alternatively stated, such an effort amounts to a synthesis of Heisenberg's uncertainty principle and Einstein's relativity—a prospect that was anathema to the great relativist. (The Heisenberg principle —some would say the very embodiment of antideterminism in the microscopic world, first enunciated by the German physicist Werner Heisenberg in the 1920s—maintains the impossibility of simultaneously measuring well *both* the position and the velocity of an elementary particle.)

The argument, more specifically, is as follows. According to relativity theory, the radius of curvature of spacetime is proportional to the mass enclosed. On the other hand, the Heisenberg principle states that no particle can be located with a precision better than to within a distance that is inversely proportional to the mass. For ordinary objects like stars, the first distance is vastly larger than the second, and there is no contradiction. As the mass becomes smaller, however, the radius of curvature decreases and the distance uncertainty increases, until they finally equal each other at a mass of roughly 10^{-5} gram; it is at this mass that both the radius of curvature and the distance uncertainty equal 10^{-33} centimeter, a distance known as the Planck length.

In a recent and potentially relevant development, researchers have become excited about a radical idea proposed nearly two decades ago. Called "superstrings," this theory aspires to unite all the laws of physics into a single mathematical framework. The name derives from the novel idea that the ultimate building blocks of nature aren't point particles at all, but tiny vibrating strings. If this view is correct, it means that the protons and neutrons in all matter, from our bodies to the farthest star, are fundamentally made of strings. However, no one has ever seen such strings, since they are predicted to be more than a billion billion times smaller than a proton—in fact, 10^{-33} centimeter, the Planck length. Depending on the mode of vibration, separate particles of matter can be created from the subatomic strings, much the way a violin string can resonate with different frequencies, each one creating a separate tune of the musical scale. Disconcertingly, the theory of superstrings works only if the Universe began with ten dimensions, six of which (somehow) became "hidden" near the time of the big bang. To some physicists, such a revolutionary idea borders on science fiction (and even theology), whereas for others it possesses breathtaking elegance. Even so, the world of science is littered with mathematically elegant theories that apparently have no basis in physical reality. And although the theory of superstrings is now causing great excitement in the physics community, there is to date not a shred of experimental or observational evidence to support it.

At any rate, our current knowledge of strong gravitational forces implies that such quantum effects very likely become important whenever the Universe is even more energetic than we have yet considered. Aside from potential black holes, which we shall soon discuss, such huge energies could have prevailed only at times earlier than 10^{-35} second, when the Universe was even hotter and denser. Specifically, at a time of 10^{-43} second, when the average temperature was about 10^{32} degrees, the four known forces are thought to have been one—a truly fundamental force operating at energies characterizing one of the earliest cosmic epochs imaginable. Only at smaller energies (that is, at times after 10^{-43} second) would the more familiar four forces begin to manifest themselves distinctly, though in reality all four are merely different aspects of the single, fundamental, super-grand force that ruled at (or near) creation.

Any theory that penetrates even closer to creation is currently hardly more than conjecture, though many researchers have a "gut feeling" that, once we have in hand the proper theory of quantum gravitation, our understanding might automatically include a *natural* description of creation itself. It is not inconceivable that the primal energy emerged at zero time from quite literally nothing, uncannily in accord with the structureless singularity described by the time-honored poetic expression ". . . without form and void, with darkness upon the face of the deep." This might be true because, even in a perfect vacuum—a region of space containing neither matter nor energy—particle-antiparticle pairs (such as an electron and its antiparticle opposite, the positron) are constantly created and annihilated in a time span too short to observe. Though it seems impossible that a particle could materialize from nothing—not even from energy—it so happens that no laws of physics are violated, because the particle is annihilated by its corresponding antiparticle before either one can be detected. Furthermore, for such events *not* to happen would violate the laws of quantum physics, which cite, via the Heisenberg principle, the impossibility of determining *exactly* the energy content of a system at every moment in time. Hence, natural fluctuations in energy content must occur, *even when the average energy present is zero.*

In this way, our Universe may well have been a case of *creatio ex nihilo,* by means of an energy change that lasted for an unimaginably short duration—a "self-creating Universe" that erupted into existence spontaneously, much as elementary particles occasionally and suddenly originate from nowhere during certain subnuclear reactions. Such a "statistical" creation of the primal cosmic energy from absolutely nothing has been somewhat sacrilegiously dubbed "the ultimate free lunch." It may indeed be the extreme manifestation of the long-standing quip "Nature abhors a vacuum." It may also be the solution to the time-honored philosophical query "Why is there something instead of nothing?" The answer, ostensibly, is that the probability is greater that "something" rather than "nothing" will happen. Clearly, the development of a quantum gravitational description of events at "$t = 0$" is the foremost challenge in the realm of physics today.

Interestingly enough, we can examine (and, we hope, test) the nature of this ultimate singularity, which presumably existed some fifteen billion years ago, by studying other, much smaller singularities thought to be present at various places in today's Universe. These smaller singularities are called black holes.

A black hole is a region containing a huge amount of mass spread throughout an extremely small volume. It's not an object per se so much as a hole, and one that is dark. These two factors —large mass and small size—guarantee an enormously strong gravitational force. Why? Because one-half of the law of gravity stipulates that gravity is directly proportional to the mass. The other half dictates that gravity is inversely proportional to the square of the distance over which the matter is spread. Thus, because the distance term is squared, the gravitational force grows spectacularly when a huge mass is compressed. Accordingly, spacetime is expected to be severely warped near black holes; a knowledge of relativity is essential when dealing with these most bizarre regions.

To place black holes into perspective, we need to examine some of the physical properties of normal stars, including white-dwarf and neutron stars that possess rather peculiar properties in their own right. The material of the next chapter will then prepare us to better appreciate the genuinely baffling features of nature's odd singularities, to be discussed further in subsequent chapters.

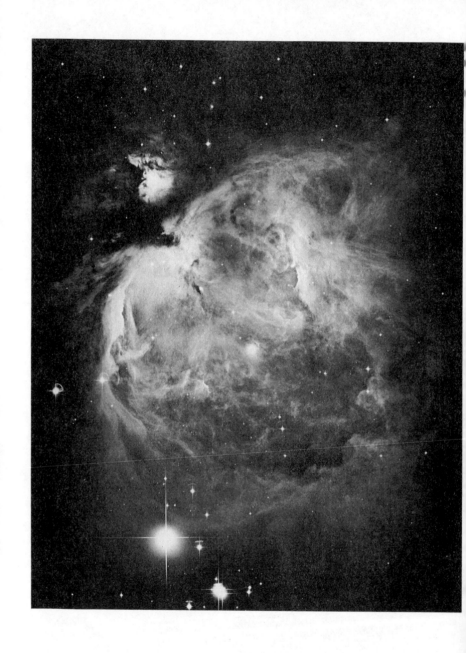

Great Nebula in Orion, a region of star formation

FIFTEEN

Highlights of Stellar Evolution

SCATTERED amid the dark realms of the nighttime sky are myriad dusty blobs of gas, each often stretching over light-year dimensions. These so-called interstellar clouds occasionally begin gravitationally infalling by accumulating matter, thereby progressively increasing their gas density, atomic collisions, and heat developed within. A protostar eventually forms, becoming a genuine star after a relatively short time, by astronomical standards (a few tens of millions of years on the average). Once fully formed, stars exist as mostly unchanging balls of gas, the inward pull of gravity fairly well balanced against the outward pressure of hot gas.

The total lifetime of a star depends largely on its mass and can vary broadly between a billion and a trillion years. Contrary to intuition, the least massive stars endure the longest (owing to their sluggish rate of nuclear fusion). But once a star begins running out of fuel, it does so rapidly, passing through a series of evolutionary stages characterized by periodic instability. Figure 39 summarizes the various paths whereby matter can enter the graveyard of old stars.

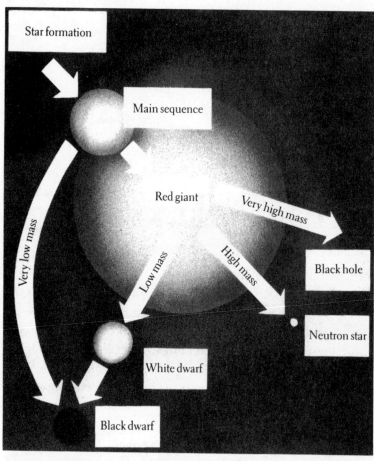

FIGURE 39 This illustration summarizes the evolutionary paths of all known stellar objects. After emerging from the placental envelope within its parent interstellar cloud, a typical (main-sequence) star passes through the red-giant phase, after which its ultimate fate depends upon its mass. Low-mass stars, of which our Sun is a member, all probably fade away as black dwarfs; high-mass stars seem destined to become ultracompact neutron stars, or even bizarre black holes.

Late evolutionary fates also depend mainly on the mass of the star. Low-mass objects pass through the swollen red-giant stage (such as the reddish Betelgeuse, in the constellation Orion), shed their outermost layers as nebular shells of gas (such as the Ring Nebula in Lyra), and eventually leave behind their cores, which in turn become white-dwarf stars having the relatively small size of a planet but the relatively large mass of a star (such as Sirius B, noted in figure 22). (Our Sun is so destined after about ten billion years of steady shining, some five billion years from now.) The gravitational pull of the matter in a white-dwarf star, so dense that a teaspoon of it would weigh a ton on Earth, is counterbalanced essentially by the pressure of the microscopic clouds of charged electrons in the star. Such stars are fated to shrivel up into black dwarfs—cold, dark cinders in the near-void of space.

High-mass objects are not nearly so well behaved. Rather than gently dispersing layers of gas as in the expulsion of a nebula, massive stars violently eject large amounts of their matter into interstellar space by means of supernova explosions (such as the Crab Nebula in Taurus). These stars, considerably more massive than our Sun, are also thought to leave at the supernova core an intact remnant known as a neutron star (such as the "pulsar" in the midst of the Crab's debris). Having the size of an average city yet a phenomenal density of a million tons per cubic centimeter, such tremendously compacted objects counterbalance gravity by the sheer impenetrability of neutrons themselves.

The theory of stellar evolution makes specific predictions about the mass of the core remnant. Should the burned-out core be less than about 1.5 times the mass of our Sun (the so-called Chandrasekhar limit, after S. Chandrasekhar, an Indian-American theorist), the resultant object is expected to become a white dwarf. If the core mass lies between 1.5 and 3 solar masses, the result is a neutron star; the added mass enables the star to crush its matter beyond atomic identity, driving the electrons into the protons and fashioning neutrons. And if the core remnant has a mass greater than about 3 solar masses, not even tightly packed neutrons can withstand the gravitational pull. Gravity wins out, and the object collapses, perhaps forever.

To illustrate these differences, consider the nearby binary-star system having the catalog name HD47129. Observed from Earth, its orbital size and period indicate that one of the two stars has a total mass of about 76 solar masses. This abnormally hot star is the most massive star known anywhere in the Galaxy. If the essence of figure 39 is correct, this star will surely explode as a supernova someday. It is destined to leave a remnant core having a mass less than its original 76 solar masses. But how much less is not known. If the future supernova explosion of this star ejects 74.5 or more of its original 76 solar masses, the core will become a white dwarf, as we noted above.

However, this argument is likely to be flawed. Supernovae probably do not eject such a high fraction of a star's original mass. A combination of theoretical arguments and recent observations of supernova remnants near our Sun implies that roughly half of the original mass is expelled to interstellar space. The exact amount is unknown, not only because the devastating effects of any explosion are hard to predict but also because the total amount of scattered debris in observed supernova remnants is difficult to estimate. In any case, adopting as a rule of thumb that about half of the original star's matter is ejected during a super-nova, we can expect that the future explosion of this 76-solar-mass star will most likely yield a remnant core of approximately 30 to 40 solar masses. Too massive to be a white dwarf, it is also too massive to be a neutron star.

The dividing lines at 1.5 and 3 solar masses are probably inaccurate because they essentially ignore the effects of magnetism and rotation, two physical properties surely present in stars. Since these effects can compete with gravity, they must influence the evolution of stars. We cannot be certain by how much the mass dividing lines will change, for no one really knows to what extent the basic laws of physics are valid in regions of very dense matter that is both spinning and magnetized. If anything, these dividing lines will shift upward when magnetism and rotation are included, because even larger amounts of mass will be needed for gravity to compress stellar cores into neutron stars or black holes.

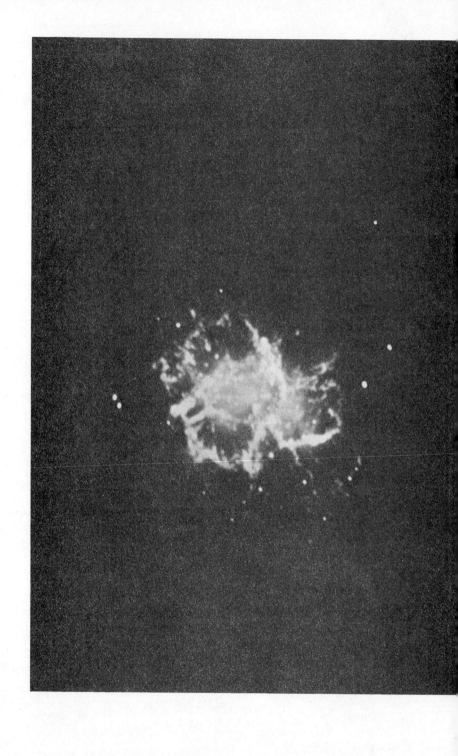

SIXTEEN

Disappearing Matter

WHAT HAPPENS when a star's remnant core exceeds three solar masses? What makes this value so special? The answer is that this is the mass value for which gravity can (apparently) no longer be countered. Not even neutrons, touching one upon another, can halt the pull of gravity within such a massive, compact, and relatively cool object. According to theory, such an object just continues falling inward, crushing matter to the dimensions of a point.

Of course, scientists might someday discover a new countervailing agent capable of withstanding the gravitational pull of a compact, massive object. But being currently unaware of any such force, we are compelled to conclude that gravity will cause a remnant core of greater than several solar masses to infall forever. An entire star—larger than our Sun—collapses to an object the size of Earth, then the size of a city, then that of a house, a human, a pinhead, a microbe, and beyond. The star catastroph-

---

The Crab Nebula in Taurus, an exploded star after death

ically implodes without limit; apparently nothing can stop it.

How can we possibly appreciate such a seemingly ridiculous phenomenon? How can an entire star shrink to the size of an elementary particle, while presumably on its way to even smaller dimensions? Does this make sense? Well, this is what the detailed mathematics predicts. Without some agent to compete against gravity, massive core remnants are expected to shrink to a singular point of infinitely small volume.

A complete analysis of the complex mathematics needed to understand the true nature of black holes is beyond the scope of this book. Nonetheless, we can usefully explore a few qualitative aspects of these incredibly dense and bizarre regions of spacetime.

Consider first of all the concept of escape velocity. This is the velocity that any small object needs to escape from a large one. The following relation shows that the escape velocity is proportional to the square root of the planet's mass divided by the square root of its radius:

$$\text{escape velocity} \propto \sqrt{\text{planet's mass} / \text{planet's radius}}$$

For example, on Earth, with a radius of about 6,500 kilometers, the escape velocity is nearly 11 kilometers/second. To launch any object—molecule, baseball, rocket, whatever—away from Earth, we must make that object move faster than this. (This explains why modern cannon projectiles, fired at some 2 kilometers/second, return to Earth's surface.)

Imagine now a hypothetical experiment for which the apparatus is a huge three-dimensional vise. In fact, let the vise be large enough to hold the entire Earth. Imagine, furthermore, that Earth is squeezed on all sides. As our planet shrinks under the onslaught, its density rises because the total amount of mass remains constant inside an ever-decreasing volume.

Suppose that our planet is compressed to one-quarter its present size. The proportionality noted above predicts that the escape

FIGURE 40 Earth being squeezed in a gigantic vise—just a thought!

velocity is then doubled. Any object attempting to escape from this hypothetically compressed Earth would need a velocity of about 22 kilometers/second.

Imagine compressing Earth still more. Squeeze it, for example, by an additional factor of 1,000, making its radius hardly more than a kilometer. Accordingly, the escape velocity increases dramatically. In fact, a velocity of roughly 700 kilometers/second is required to escape from an object having a radius of about a kilometer and a mass of about all of Earth.

This, then, is the trend. As an object of any mass contracts, the gravitational force grows stronger at its surface, mostly because of increasing density.

If we could imagine Earth being further compacted, the escape velocity would rise accordingly. In fact, if our hypothetical vise were to squeeze Earth hard enough to crush its radius to about a centimeter, the velocity needed to escape its surface would reach 300,000 kilometers/second. Do not gloss over the number; this is no ordinary velocity. It is the velocity of light, the greatest velocity allowed by the laws of physics, as we now know them.

So if, by some fantastic means, the entire planet Earth could be compressed to the size of a pea, then the velocity for anything to escape would have to exceed that of light. And since that seems impossible, the compelling conclusion is that nothing—absolutely nothing—can escape from the surface of such a compressed "Earth."

Thus, if our planet were squeezed to less than centimeter dimensions, we could legitimately argue that knowledge of it would be lost to the rest of the Universe. After all, there would be no way for a launched rocket, beam of light, or any type of radiation to get away. Such a compressed object would have become invisible and uncommunicative, for no information whatsoever could be exchanged with the Universe beyond. For most practical purposes, such a supercompact object can be said to have disappeared from the Universe!

The above example is, of course, hypothetical. Nothing in the known Universe resembles a vise capable of squeezing the entire Earth to centimeter dimensions. But in massive stars, such a vise does in fact exist. It's known as gravity.

Gravity cannot crush Earth in this way, because our planet simply harbors too little mass; the collective gravitational pull of every part of Earth on all other parts of Earth is just not powerful enough. However, at the end of a star's life, when the nuclear fires have dwindled, gravity can literally crush a star on all sides, thereby packing a vast amount of matter into a very small sphere.

When stellar core remnants house more than three solar

masses, the critical size at which the escape velocity equals that of light is not, as for Earth, of centimeter dimensions. For typically massive remnants, this critical size is on the order of kilometers. For example, a ten-solar-mass remnant would have a critical size of nearly thirty kilometers. To be sure, it is no less a feat to compress an entire star to kilometer dimensions than to compress a planet-sized object to centimeter size. In the stellar case, however, we are not talking about a hypothetical situation using an imaginary vise. The relentless pull of gravity is strong enough to compress spent stars to extraordinarily small dimensions. The strong self-gravity of massive stars is real.

Astronomers have a special name for the critical size below which any object is predicted to disappear in this manner. Called the "event horizon" (or, technically, the "Schwarzschild radius," after K. Schwarzschild, a German theorist), this size defines the region within which no event can ever be seen, heard, or known by anyone outside. Accordingly, the event horizons of Earth and of a ten-solar-mass star are one centimeter and thirty kilometers, respectively. (A quick rule of thumb states that the size of the event horizon equals three kilometers multiplied by the object's mass, provided that mass is expressed in units of solar masses.)

So, we might say that magicians could in fact make coins and other small objects disappear, assuming they squeezed their hands hard enough. Even people could disappear if they could arrange to be compressed to a size smaller than 10^{-23} centimeter. Gravity will not do it to us, though. Humans are just not massive enough. The collective gravitational pull of all the atoms in our bodies falls far short of the force needed to compress us to this minuscule size. Nor are there any technological means currently known that even come close to doing so—not even modern garbage compactors. Luckily for us!

On the other hand, perhaps someday, through some marvel of technology, our descendants will learn how to compact garbage to an almost incredibly small size. It would then disappear! Maybe some practical application will eventually result from black-hole research after all.

The important point here is the following: if no force is capable of withstanding the self-gravity of a dead star having more than three solar masses, such an object will naturally collapse of its own accord to smaller and smaller dimensions. Such a stellar core remnant will not even stop infalling at its event horizon. An event horizon is not a physical boundary of any type, just a communications barrier. The core remnant shrinks right on past it, to ever-diminishing sizes, presumably on its way toward becoming an infinitely small point—a singularity. We say "presumably" because we cannot absolutely rule out as-yet-undiscovered forces capable of halting catastrophic collapse somewhere between the event horizon and the point of singularity. The subject needed to reveal any such force—quantum gravity—has yet to be invented.

Here, then, is the observational sequence of events if these late stages of stellar evolution are correct. A very massive star ends its burning cycle by exploding as a supernova. Approximately half the star's original content is then ejected as fast-moving debris. Provided at least three solar masses remain behind, the unexploded remnant core will collapse catastrophically, the whole core diving below the event horizon in less than a second. The core simply winks out—not merely becoming invisible but literally disappearing—leaving a small dark region from which nothing can escape. This is the way black holes are born, as blackened domains of space. They are not really objects as much as holes —black holes in the fabric of spacetime.

Einstein with Hideki Yukawa and John Archibald Wheeler, Princeton, 1954

SEVENTEEN

Properties of Black Holes

MODERN NOTIONS about black holes rest solely on the theory of general relativity. Whereas white-dwarf and neutron stars are valid end points of stellar evolution within the confines of the Newtonian theory of gravity, only the Einsteinian theory of spacetime specifies the physical properties of bizarre phenomena like black holes (though such peculiar objects were broadly anticipated as early as the late eighteenth century by the Englishman John Michell and by the French mathematician Pierre Simon de Laplace, who used classical escape velocity arguments like those of the preceding chapter). As predictions of relativity theory, black holes should obey all its standard laws. In particular, the mass contained in a black hole is expected to warp space and time in its vicinity. Close to the hole, the gravitational force becomes overwhelming and the curvature of spacetime extreme. At the event horizon itself, the curvature is so great that spacetime folds over on itself, causing trapped objects to disappear.

Several props can help us visualize the curvature of spacetime near a black hole. Each way is, however, only an analogy; it is not really an example. The problem here, as earlier in this book, is our inability to work conceptually in four dimensions.

The formation of black holes and the extreme warping of spacetime caused by them can nonetheless be appreciated by considering the analogy of a large family of people living on an enormous rubber sheet—a sort of gigantic trampoline. Deciding to hold a reunion, they converge on a given place at a given time. Like any reunion, this is an event in spacetime. However, as is shown in figure 41, one person remains behind, wishing not to attend the gathering; he can keep in touch by means of message balls rolled out to him along the rubber sheet. These balls are the analogue of radiation traveling at the velocity of light, while the rubber sheet mimics the fabric of spacetime itself.

As the people converge, the rubber sheet begins to sag more and more. Their accumulating mass in a small place creates an increasing amount of spacetime curvature. The message balls can still reach the lone person far away in essentially flat spacetime, but they arrive less frequently as the sheet becomes progressively more warped, as in figure 41 (*b* and *c*).

Finally, when enough people have arrived at the appointed spot, the mass becomes too great for the rubber to support. As is illustrated in figure 41 *(d)*, the sheet closes off into a bubble, which compresses the people into oblivion.

FIGURE 41 Any mass, even the small amount contained in people, causes spacetime to be curved. To appreciate this figure, imagine numerous people frolicking on a gigantic trampoline made of thin rubber membrane *(a)*. As all those except a single person converge toward a given place, the curvature of the sheet grows progressively larger *(b)* and *(c)*. The rapidly sinking crowd can still communicate with the distant loner by, say, rolling message balls to him. Finally, as the people assemble at a particular spot, their combined mass is enough to drag them inside a sealed-off bubble *(d)*, forever trapping them and their message balls.

Until the end, message balls were able to reach the lone survivor, albeit at a lower and lower rate. But as the rubber sheet segregates the bubble of people (thereby forming an event horizon), the balls can no longer return to the person left behind. Regardless of the speed of the last message ball, it cannot quite outrun the downward-stretching sheet.

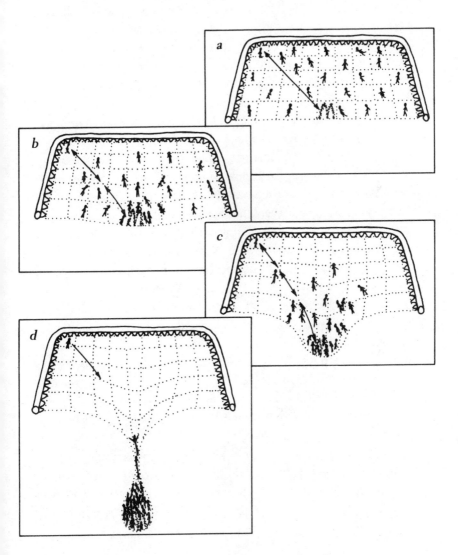

This analogy (very) roughly depicts how a black hole is predicted to warp spacetime completely around on itself, thus isolating it from the rest of the Universe.

Be sure to note a basic feature of black holes: they are not cosmic vacuum cleaners; they do not cruise around interstellar space, sucking up everything in sight. The movements of objects near black holes mimic those of any objects near a very massive star. The only difference is that, in the case of a black hole, objects skirt or orbit about a dark, invisible region. Neither emitted nor reflected radiation of any sort emanates from the position of the black hole itself.

Black holes, then, do not go out of their way to drag in matter, but if some matter does happen to fall in through the normal pull of gravity, it will be unable to get out. Black holes are like turnstiles, permitting matter to flow in only one direction—inward. Swallowing matter, they constantly increase their mass; their size also grows, since the event horizon depends on the amount of mass trapped inside. If black holes really do exist in space, all of them are probably growing in mass and size, as increasing quantities of matter accumulate below their event horizons.

Another important point about the nature of black holes is this: matter flowing into a black hole is subject to great tidal stress. An unfortunate person, falling feet first into such a hole, would find himself stretched enormously in height, all the while being squeezed laterally. He would be literally torn apart, for gravity would be stronger at his feet than at his head. He would not stay in one piece for more than a fraction of a second after passing the event horizon.

Any kind of matter near a black hole undergoes similar distortion and breakup. Whatever falls in—gas, people, space probes—is vertically elongated and horizontally compressed, in the process being accelerated to high speeds. The upshot is numerous and violent collisions between the torn-up debris, all of which yield a great deal of heating (by friction) of the infalling matter.

The rapid heating of matter by tides and collisions is so efficient that, prior to submersion below the hole's event horizon, newly infalling matter emits radiation on its own accord. This is

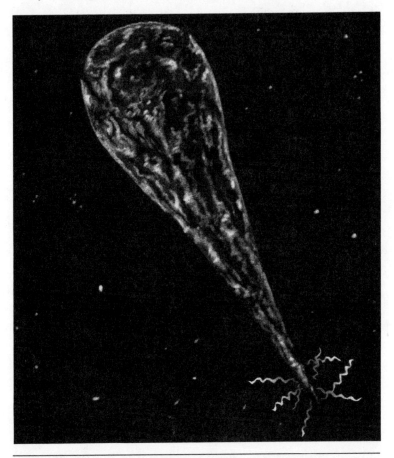

FIGURE 42 Any matter falling into the clutches of a black hole will become severely distorted and thus heated by friction. In this sketch, an imaginary planet is being pulled apart by the gravitational tides of the black hole. The "lightning bolts" near the hole are meant to depict the emitted X rays from a region just outside the hole's event horizon.

simple thermal radiation (the kind released by stars, light bulbs, and so on), emitted because the infalling matter has become so hot that the radiation is expected to be of the X-ray type. In effect, the gravitational potential energy of matter outside the black hole is converted to heat energy while the matter falls toward the hole. Once the hot matter dips below the event horizon, the radiation ceases to be detectable.

Contrary to popular belief, then, black-hole environments are expected to be *sources* of energy. If the amount of infalling matter is large, regions just outside event horizons could be absolutely prodigious emitters of intense radiation. The ability technologically to tap this energy might well be a major milestone in the history of long-lived civilizations.

Before leaving this chapter on the properties of black holes, a significant caveat deserves mention. The subatomic properties of matter are an important feature so far neglected in theories of black holes. This could be a crucial issue because all the matter trapped within a black hole is predicted to collapse to the size of an elementary particle or less. To decipher the nature of matter deep down inside the event horizon, scientists will have to merge their knowledge of gravitational physics with that of subatomic physics—the yet-to-be-invented subject of "quantum gravity," discussed briefly in chapter 14.

Some attempts to understand gravity on a microscopic scale suggest that black holes may not be entirely black after all. When subatomic physics is taken into account, it seems that matter and radiation might be able to escape from black holes. This might be true because when the phenomenon of pair creation (also noted in chapter 14) occurs near a black hole, one particle could conceivably fall into the black hole before the pair of particles had a chance to annihilate. The other particle would then be free to leave the scene, making the black hole appear to the outside world as a source of radiation. According to these preliminary ideas, black holes might not last forever. They slowly evaporate (or dissipate their matter and energy) and finally explode, scattering their contents into interstellar space.

Just like that of ordinary stars, the lifetime of a black hole depends on its mass. A conventional black hole having several times the mass of the Sun is predicted to explode after many times the current age of the Universe. Thus, the issue is moot for the usual type of black hole described elsewhere in this chapter; astronomers cannot expect to observe either their slow decay or their ultimate explosion.

These new theories, however, also predict that the pressure in the early Universe may have been just right to compress pockets of matter into miniature black holes. For example, very small black holes having a mass of about 10^{15} grams are predicted to have a lifetime nearly equal to the current age of the Universe. (The Sun's mass is about 10^{33} grams.) Such a mini–black hole, having the mass of a typical meteoroid or a terrestrial mountain, would have an event horizon about equal to that of a subatomic particle ($\cong 10^{-13}$ centimeter). Provided they do exist, such mini–black holes should be exploding now and thus emitting intense bursts of gamma-ray radiation.

Although the theory remains to be proved, if astronomers detect this radiation—and they are are now trying to do so—it will conceivably tell us something not only about the nature of black holes but also about the physical conditions prevalent in the immediate aftermath of the big bang.

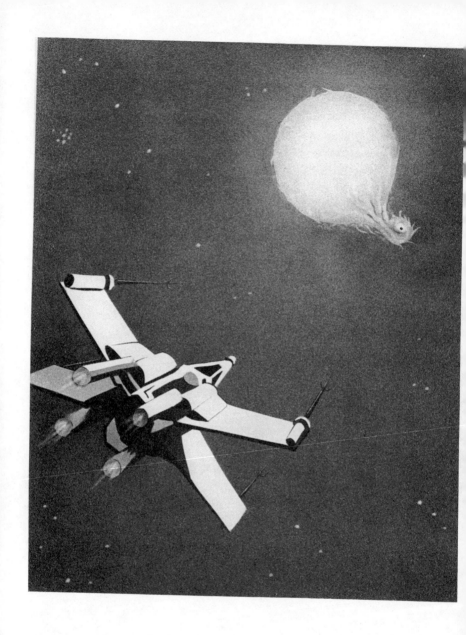

One-way Ticket

EIGHTEEN

Space Travel Near Black Holes

C ONSIDER the prospects of exploring a black hole. It would be wise to conduct such studies from a safe distance, for travelers approaching too close may well find themselves pulled past the event horizon by the hole's gravity—never to return.

Could we voyage anywhere near a black hole then? One reasonably safe way to study a black hole up close would be to carefully orbit it. A stable, circular orbit could get us quite close, even though we would be immersed in the hole's strong gravitational force. After all, Earth and the other planets of our Solar System each orbit the Sun without falling into it; the Sun's gravitational pull is offset by each planet's forward momentum, as was described in chapter 3. The gravity field around a black hole is basically no different, albeit stronger, and should permit stable orbits.

Even so, travel would be unsafe for humans, should they

venture (even in a stable orbit) too close to the hole. The extrapolation of human endurance tests conducted on American and Russian astronauts suggests that the human body cannot withstand physiological stress more than ten times the present gravitational pull we normally feel on Earth's surface (that is, ten g's). This more stressful state would occur at about 3,000 kilometers from a ten-solar-mass black hole (which, recall, would have a 30-kilometer event horizon). At points closer than that, gravity would probably tear the human body apart.

We need not send a human to study a black hole. Automated space probes could be designed to withstand stressful conditions normally intolerable to humans. For purposes of discussion, let us use only indestructible astronauts, mechanical robots of sorts. To make it interesting, let us send these hypothetical robots toward the center of the hole. Watching from a distance in our safely orbiting spacecraft, we shall then be able to test the nature of space and time near the hole. In this way, the mechanical robot becomes a test particle from whose behavior we can infer several things about black holes.

Dispatching robot spacecraft may in fact be the only way we can ever study black holes (nearly) in situ. Such robots could be useful probes of theoretical ideas, at least down to the event horizon. After that, there is no known way for a probe to return any information about its findings.

Suppose, for example, that the robot has an accurate clock and a light source mounted on its side. Distant humans, safely orbiting far from the event horizon, would be able to use telescopes to read the clock and measure the light. The information received should help us decipher the nature of spacetime in the vicinity of the hole as the robot astronaut approaches the hole. What might we discover?

First of all, recall our discussion in chapter 7 concerning the behavior of light (or any type of radiation) in a strong gravitational

FIGURE 43 Robots could travel toward a black hole while performing experiments that humans, farther away, can monitor in order to learn much about the nature of spacetime near such a hole. Here, a spaceship in a safe, stable orbit has deployed a robot probe to spiral ever closer to the black hole. Once the robot passes the horizon and enters the hole's innards, however, the experiment ends—at least from the perspective of anyone outside the hole, for no information whatever can escape the unimaginably warped blackness.

force field. Photons expend some of their own energy while moving away from the source of gravity. The photons lose energy because they must work to get away. They do not slow down at all; they just lose energy. And since a photon's energy is proportional to the frequency of its radiation, light that loses energy will have its frequency reduced (or, conversely, its wavelength increased). Hence, radiation emitted from the vicinity of a black hole will be red shifted by an amount depending upon the strength of the gravitational force. This is not a red shift caused by motion (Doppler effect); rather, it is a red shift induced by gravity (or spacetime curvature).

Most objects' gravity is insufficient to shift radiation measurably toward the red. Sunlight, for example, is not red shifted by any detectable amount. As we noted in chapter 7, a few white-dwarf stars do show a slight reddening of their emitted light. And neutron stars should appreciably shift their radiation, but it is currently impossible to disentangle the effects of gravity and magnetism on the observed signals. Only near black holes is the gravitational pull so great that a red shift should be unambiguously measurable, at least in principle.

So as photons travel from the robot's light source to the orbiting spacecraft, they become gravitationally red shifted. From the viewpoint of the safely orbiting humans, a green light, for instance, would progressively become yellow and then red as the robot astronaut neared the black hole.

As the robot approached the event horizon, radiation from its light source would become undetectable with optical telescopes. The radiation reaching the humans in the orbiting spacecraft would, by then, be lengthened so much that infrared and radio telescopes would be needed to receive it. Light trying to escape from an infinitesimal location just above the event horizon is expected to become gravitationally red shifted to nearly infinitely long wavelengths. In other words, light uses virtually all its original energy in trying to escape from the edge of the hole. What was once visible light (on the robot) has hardly any energy left upon arrival at the safely orbiting spacecraft. Theoretically, this radiation reaches us—still moving at the velocity of light, in fact—but with virtually zero energy. The originally emitted light radiation has become red shifted to wavelengths longer than conventional radio waves, indeed beyond our perception. Neither humans nor our equipment could detect it.

What about the robot's clock? Assuming that the distant observers in the safely orbiting spacecraft can read it, what time does it tell? Is there any observable change in the rate at which the clock ticks while moving deeper into the gravitational force field?

Einstein's theory of relativity (see chapters 5 and 6) suggests that, from the viewpoint of the safely orbiting spacecraft, any clock close to the hole would operate more slowly than an equivalent clock on board the spacecraft. The clock closest to the hole would operate slowest of all. Upon reaching the event horizon, the clock would stop altogether. It would be as if the robot astronaut had achieved immortality! All action would become virtually frozen in time. Consequently, an external observer could never really witness an infalling astronaut sink below the event horizon. Such a process would take forever.

However, relativity theory predicts that, from the viewpoint of the infalling robot, there will be no strange effects at all. To the computer on board the indestructible robot, the light source has not reddened and the clock keeps perfect time. In the frame of reference of the infalling robot, everything is normal.

Nothing prohibits the robot traveler from passing right through the event horizon of a black hole. Neither the laws of physics nor those of biology physiologically constrain objects from moving ever closer to the black-hole singularity. Nothing resembling a brick wall lurks at the event horizon; it's only an imaginary boundary in space. Travelers passing through might not even know it—which undoubtedly becomes a problem, for they are not going to get out! Once they are inside the event horizon, only velocities greater than that of light can ensure their escape. And since that is impossible, any and all things—spaceships, people, light, information of any sort—become trapped.

No doubt, you are wondering what lies within the event horizon of a black hole. The answer is simple: no one knows.

Some researchers suggest that the inner workings of black holes are irrelevant. Experiments could conceivably be done by robots sent "down under" to test the nature of space and time inside an event horizon, but that information could never reach us on the outside. Apparently, theories of the recesses of black holes cannot be put to the experimental test. Anyone's theory is as valid as anyone else's.

Perhaps the inner sanctums of black holes represent the ultimate unknowable. For that very reason, though, other researchers argue that it is of utmost importance to unravel the nature of black holes, lest we someday begin to worship them. Large segments of humankind have often revered the unknowable, venerating that which cannot be tested experimentally. Many still do.

What sense are we to make of black holes? Do all these outlandish phenomena really occur? The basis for these weird predictions is the relativistic concept that mass curves spacetime —a circumstance already tested to be a surprisingly good approximation of reality (see chapter 7). The larger the mass concentration, the greater the warp and, thus, the stranger the observational consequences. Perhaps.

Some researchers argue that relativity is incorrect, or at least incomplete, when applied to black holes. Admittedly, it seems nonsensical to claim that very massive astronomical objects will collapse catastrophically to infinitely small points. Not even the wildest imaginations can visualize this phenomenon. Perhaps the current laws of physics are inadequate in the vicinity of a singularity. In fact, we are certain that, *at* the point of singularity, relativity is incomplete and probably absurd. On the other hand, maybe matter trapped in black holes never does reach a singularity. Perhaps matter just approaches this most bizarre state, in which case relativity theory may still hold true. Scientists just do not yet know.

NINETEEN

Observational Evidence

THEORETICAL ideas aside, let us examine how black holes might be observed. After all, if they cannot be detected observationally, they might not even exist. Perhaps the computer modeling of the final stages of stellar evolution is incorrect in predicting that massive stars spawn remnant cores larger than three solar masses. Maybe all the original star is blasted to smithereens. Or perhaps yet another, undiscovered force exists capable of competing with gravity despite the extreme conditions of ultracondensed matter expected in black holes.

Is there any observational evidence for black holes? More to the point, do we have methods of probing such invisible objects populating the depths of space?

Although black holes are invisible, they are expected to remain strong sources of gravity. Accordingly, astronomers should be able to test for a hole's existence by studying its associated

The *Einstein Observatory*, now an orbiting spacecraft, being readied for launch gravitational force (or spacetime curvature). For example, the motion of a spacecraft could conceivably be used to probe the outer environment of a black hole. Any craft should behave dynamically just as though there were a massive, visible object at the site of the hole. In other words, most conventional means of assessing a black hole disappear, but its gravity persists.

Thus, we could, at least in principle, maneuver a spacecraft into an orbit at a safe distance outside a region of otherwise total darkness. Somewhat as in the case discussed in chapter 18 and sketched in figure 43, nothing would necessarily be perceived at the orbit's focus, but the hole's gravity would control the space-craft's motion such as the Sun guides Earth in its orbit. The spacecraft and the black hole then form a truly odd couple, a peculiar version of a binary-star system wherein two objects orbit about their common center of mass.

With the spacecraft safely in position, the law of gravity could be used to infer the hole's mass. Knowing the orbital size and period of the spacecraft about the hole, we could compute (by means of one of Kepler's laws) the combined mass of the hole and spacecraft. And since the spacecraft's mass will surely be negligible compared with the black hole's, the orbital properties thus directly yield the mass of the black hole. In this way, we can infer the existence and mass of a black hole without actually observing anything deep down inside it.

Another way that black holes might be detected, again at least in principle, also involves a binary system (though in actuality a binary system is not strictly necessary). This test requires us to observe a small black hole while it passes in front of a much larger and visible companion. The observable effect we might expect would be a minute dot of blackness gliding across an otherwise bright star. But this would be extremely hard to see; the 12,000-kilometer planet Venus is barely noticeable when it transits the Sun, so a kilometer-sized object moving across the image of a faraway star would be virtually invisible.

This test is not even as clear-cut as just described. The empirical effect would not really be a black dot superposed on a bright background. The problem is more complex because the background starlight would be deflected while grazing the edges of the black hole on its way to Earth. The effect resembles the bending of distant starlight around the edge of the Sun, as was described in chapter 7. In the case of the Sun, the bending amounts to less than a couple of arc seconds, an extremely small angular deflection. But since this angle is proportional (approximately) to the density of the deflecting object, the deflection of radiation should be much larger near small, massive black holes. So, our perception of a black hole in front of a bright companion star would not show a neat, well-defined black dot, however small. Instead, the bending of light around all sides of the hole would fuzz its image, thus diminishing its contrast against the background star. Recent studies have shown that such a blur would be virtually impossible to observe either with modern telescopes or with any equipment we are likely to have in the foreseeable future.

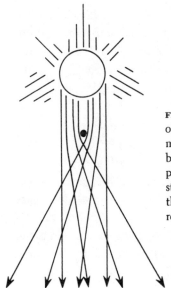

FIGURE 44 The strong gravitational bending of light around the edges of a minute yet massive black hole makes it virtually impossible to observe the hole as a black dot superposed against the bright background of any stellar companion. The resulting image of the hole would very likely be fuzzed beyond recognition.

The upshot is that these two methods of identifying and studying black holes are fine in principle but not in practice. Our civilization does not have the capability to maneuver spacecraft into the neighborhood of suspected black holes, even if we knew their exact locations. Nor do we have telescopes sophisticated enough to detect the fuzzy and diluted images of black holes passing in front of bright stars.

The idea that some black holes might be members of natural binary-star systems leads to a promising indirect test of their existence. Although black holes cannot be seen directly, they are sure to block some light from their visible companions. Once again, as happened at the turn of the century when they were the rage among observational astronomers, variable stars—those stars that regularly change their brightness—have become a fashionable and important subject in astronomy.

Figure 45 shows a light curve of a typical binary-star system. Data of this sort can yield the mass of the system, provided its orbital period and size are known. We need not see each star directly; we need only observe the regular light variations of one star to infer both the existence of an unseen companion and some of the inherent properties of the invisible object. (Measurements of the periodic Doppler shift of only one star of a binary system are also enough for us to infer the existence and some properties of an invisible companion.)

The Milky Way Galaxy harbors many such binary-star systems for which only one object can be seen. In the majority of cases, the invisible companion is probably small and dim, nothing more than a low-temperature star hidden in the glare of a much hotter star partner. In other cases, dust or other interstellar debris probably shrouds one object, making it seem invisible even when using the best equipment of our high-tech society. In either kind of situation, the invisible object is not a black hole.

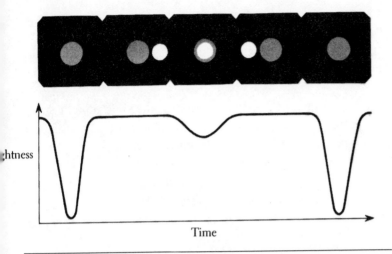

Brightness

Time

FIGURE 45 The "light curve" of a typical binary-star system displays characteristic changes in the detected light. Illustrated here is the case of the eclipsing binary Algol, the Demon Star (also known technically as Beta Persei). As drawn in the five frames *(top)*, the brighter component of this binary—a hot, main-sequence star—is periodically eclipsed by a larger, cooler star too faint to be observed separately. Every 2.9 days, as the left and right frames depict, the brighter star is fully obscured by the larger companion, causing a considerable decrease in the observed brightness, as is shown by the light curve *(bottom)*.

However, a few binary systems have peculiarities implying that one of their members is indeed a black hole. Each system has been discovered and studied during the 1970s and 1980s by Earth-orbiting satellites, for only space-based apparatus can detect the large amounts of emitted X-ray radiation. This high-frequency radiation cannot easily penetrate dust, making it unlikely that galactic debris has camouflaged one of the partners. Furthermore, in a few cases, the mass inferred for the X ray emitting invisible object equals several solar masses, thus effectively ruling out small, dim stars. Here's a description of one such binary system, considered by many researchers as the best case for a "classical" black hole.

Figure 46 shows the area of the sky known to the ancients as Cygnus (the Swan), a name still used today. The box denotes the celestial system of interest, some 7,000 light-years from Earth. The main features of many observations of this system are as follows:

1. Spectroscopic (Doppler) studies of the emitted light radiation show that the bright object—a blue-giant star with the catalog name of HDE226868—is only one member of a binary-star system whose orbital period (5.6 days) and size (twenty million kilometers across) are well determined.

2. Other spectral observations suggest that hot gas is flowing from the bright star toward an unseen companion, called Cygnus X-1.

3. This invisible object has a mass of between five and ten

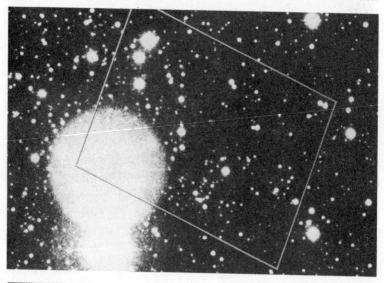

FIGURE 46 The brightest star in this photograph (cataloged as HDE226868) is a member of a binary system whose unseen companion, called Cygnus X-1, is a leading candidate for a black hole. The superposed rectangle frames both the region where the hidden companion is thought to be lurking and the field of view displayed in the next figure.

solar masses, as derived from knowledge of the binary system's orbital size and period.

4. X-ray radiation, emitted from the immediate neighborhood of Cygnus X-1, suggests the presence of scalding gas, perhaps as hot as a billion degrees Celsius.

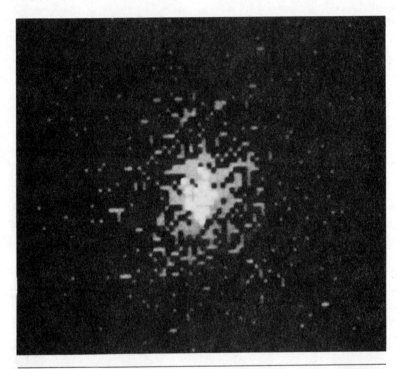

FIGURE 47 Invisible X rays, emitted from the darkness close to the center of both this figure and of the superposed rectangle of the preceding figure, define the intriguing Cygnus X-1 source. This image was obtained via the High Energy Astronomical Observatory (HEAO-B), in orbit above Earth's atmosphere, which is opaque to X rays. This spacecraft was renamed the *Einstein Observatory*, as the craft first went into operation during 1979, the centennial year of the great relativist's birth. The source's radiation can be analyzed by changing the detected X rays into radio waves (which the spacecraft does) and then telemetering those waves through the atmosphere to ground stations where they are, in turn, converted into electronic signals viewable on a computer video screen (from which this figure was taken).

5. Rapid, irregular flickering of this X-ray radiation implies that the size of Cygnus X-1 itself must be less than a hundred kilometers across.

These observational findings suggest that the invisible X ray emitting companion might well be a black hole. But since invisible black holes cannot be seen or imaged in any way, they cannot be readily illustrated either; even so, figure 48 is an artist's conception of this intriguing Cygnus X-1 region. Much of the gas drawn from the visible star (which is probably evolving toward the red-giant stage) is depicted to end up in a Life Saver–shaped disk of matter. Some of this gas inevitably streams toward the black hole,

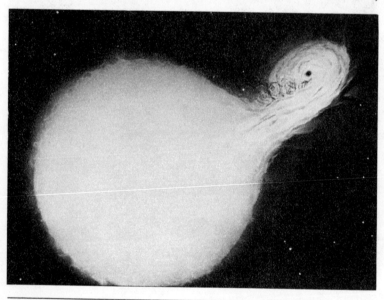

FIGURE 48 An artist's conception of Cygnus X-1 and its environment. As painted, this binary system contains a large, bright, visible star *(left)* and an invisible black hole (upper right) surrounded by a thin disk of accreted, X ray–emitting matter torn from the giant star. As with the other artwork in this book, accuracy and realism are paramount; here, for example, the disposition of the expected turbulence caused by matter in the accretion disk colliding with newly stolen gas to the lower left of the suspected hole was airbrushed in consultation with several leading black-hole theorists.

becoming superheated and X-ray emitting just before entrapment below the event horizon.

A few other similar candidates for black holes have been proposed in recent years. For example, the third discovered X-ray source in the Large Magellanic Cloud—called LMC X-3—is such an invisible object. And like Cygnus X-1, it seems to house nearly ten solar masses. In this case, LMC X-3's visible companion star seems to be distorted into the shape of an egg, apparently because of the intense gravitational attraction of the suspected black hole.

A potential problem plagues these interpretations, however. Cygnus X-1 and a few other suspected black holes in binary systems like it have masses close to the neutron star–black hole dividing line. When the effects of rotation and magnetism are someday fully included in the relativistic theory of stellar evolution, there is a chance that the dark objects in question will turn out to be more mundane. In short, they may be merely dim and dense neutron stars, and not black holes at all.

Many researchers argue that other kinds of regions show better evidence for black-hole candidates than do some of the binary-star systems just described. Current exploration of the center of the Milky Way Galaxy, some 30,000 light-years from Earth, is especially interesting in this regard. Although the core of our Galaxy is totally obscured by interstellar dust and thus cannot be studied with optical telescopes, radio and infrared observations of the innermost few hundred light-years have yielded spectacularly unexpected results. Data obtained during the past half a dozen years imply the presence of rapidly rotating hot gas, suggestive of a colossal whirlpool at the very heart of our Galaxy.

Much of the mystery surrounding the galactic center results from our inability to see it. The dark dust clouds within a few thousand light-years of our Sun effectively shroud what otherwise would be a stunning view of billions upon billions of stars tightly concentrated in and around the bulge of our Galaxy's midsection. Figure 49 *(a)* shows an (optical) view of the region of the Milky Way toward which the galactic center resides. This general direction is often referred to as the Sagittarius region, so named for one of the constellations in that part of the sky.

By employing long-wavelength techniques (especially by capturing infrared and radio radiation that is virtually unaffected by galactic debris), we have recently begun to perceive the central regions of our Galaxy. Infrared observations indicate that the galactic-core environment harbors roughly a thousand stars/cubic light-year—a stellar density well over a million times that in our solar neighborhood. Had any planets been associated with these galactic-center stars, they would doubtless have been rapidly ripped from their orbits and obliterated because the stars must often experience close encounters and even collisions. Infrared radiation also arises from what seems to be a hierarchy of warm clouds rich in dust in the central regions.

Radio observations provide additional information, especially about relatively cool gas. Figure 49 *(b)* depicts contours of (invisible) radio emission detected within the boxed portion of the optical photograph in figure 49 *(a)*. This radio radiation is the combined emission from clouds, nebulae, and myriad sites of loose galactic gas. Similar maps, not shown, have been made of neutral hydrogen, carbon monoxide, and other molecular emission in this same region.

FIGURE 49 Frame *a* shows a time-exposed photograph of stellar and interstellar matter in the direction of the constellation Sagittarius; the Milky Way's center is located within the superposed box, but because of heavy obscuration (caused largely by interstellar dust), even the largest optical telescopes can "see" no farther than one-tenth of the distance to the center. Frame *b* is a contour map of radio signals emitted by matter in the immediate vicinity of the Galaxy's center; the long-wavelength radio emission penetrates the galactic dust, providing an image of the matter in the core of the Milky Way, some 30,000 light-years away.

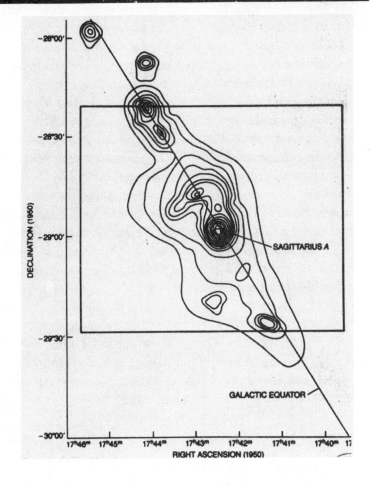

SAGITTARIUS A

GALACTIC EQUATOR

DECLINATION (1950)

-26°00'

-26°30'

-29°00'

-29°30'

-30°00'

RIGHT ASCENSION (1950)

17ʰ46ᵐ   17ʰ45ᵐ   17ʰ44ᵐ   17ʰ43ᵐ   17ʰ42ᵐ   17ʰ41ᵐ   17ʰ40ᵐ   17

Is the galactic center's mystery beginning to unravel by the decoding of radio and infrared radiation, or is our understanding of this most intriguing galactic locale still plagued by unexplained phenomena? The answer is affirmative on both counts. Astronomers now have a fairly good "road map" of the innermost regions but have yet to understand the precise mechanisms at work there.

Figure 50 places a simplified interpretation of our findings into perspective. Here a series of six airbrush conceptions capture the salient results of recent radio and infrared probings of the Milky Way's heart. Each painting centers on the Galaxy's core, and each increases in resolution by a power of ten.

Frame *a*, a reproduction of figure 1 and drawn from a viewpoint considerably beyond the Magellanic Clouds, renders the full Galaxy's morphology—its spiral arms of stars, gas, and dust, its compact, bright nucleus, its extensive, invisible halo. The scale of this initial mental image measures some 300,000 light-years across. Frame *b*, still centered on the core but magnified by an order of magnitude, spans 30,000 light-years and is highlighted by the great circular sweep of the innermost spiral arm. Moving another ten times closer, frame *c* depicts a ring of matter made mostly of giant molecular clouds and gaseous nebulae, each a few light-years in average size; this entire flattened, circular feature, about a thousand light-years in diameter, rotates rapidly ($\simeq 100$ kilometers/second) and is also known to be expanding (at a comparable velocity) away from the center—much like a cosmic version of a smoke ring released from the mouth of a successful, cigar-puffing businessman. In frame *d*, showing a region now some 300 light-years across, a pinkish ionized gas (termed plasma) surrounds the reddish heart of the Galaxy. (Despite their invisibility, these regions' colors can be inferred from their known temperatures and densities.) The source of energy producing this vast cloud of plasma is currently unknown, as is the case for the expanding ring in the preceding frame, though the two may be related. Frame *e*, spanning 30 light-years, depicts a tilted, spinning whirlpool of hot (10,000-degree) gas that marks the core environment of the Galaxy. The innermost sanctum of this gigantic whirlpool dominates frame *f*, where a swiftly spinning, white-hot disk of superheated (1,000,000-degree) gas nearly engulfs a hugely massive object that is, at the same time, too small in size to be pictured (even as a minute dot) on this, the finest of our six artist's impressions.

According to the current consensus, the galactic center is an explosive region—and one quite unlike anything discovered anywhere else in our Galaxy. The best evidence for the explosiveness is the ring of matter noted in frame *c*. On the basis of the observed expansion velocity of that ring, we can surmise that a major explosion probably occurred at or near the Galaxy's precise center some ten million years ago. The violence must have been considerable since estimates for the amount of matter now in the ring average a hundred million solar masses. Other suggestive, though less well documented, observational evidence for additional rings of matter at differing distances implies that titanic explosions might be a regular phenomenon at the center of our Galaxy.

The problem, or the mystery, concerns the cause of the explosions. Whence might the explosive energy arise? What is the source of the violence capable of sending a *hundred million* solar masses hurtling outward toward the galactic suburbs? We can only speculate about the answers at this time, but a leading contender is a supermassive black hole millions of times more massive than our Sun. Not that the hole itself need be emitting matter and energy (lest it violate the best ideas about black holes and relativity theory as well, much as we discussed in earlier chapters). Rather, the vast disk or doughnut-shaped region of matter being drawn toward the hole might regularly become gravitationally unstable as infalling matter accumulates, possibly causing periodic

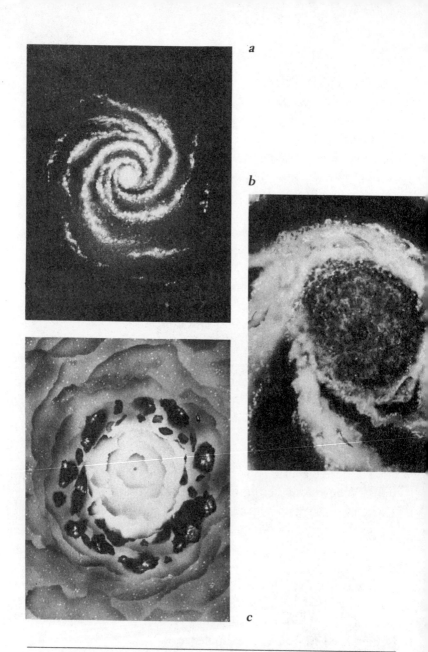

FIGURE 50 Six airbrush conceptions, each centered on the galactic core and each increasing in resolution by a power of ten. Frame *a* is a "bird's-eye view" some 300,000 light-years above the disk of the Galaxy (and is a reproduction of fig-

ure 1). Frame *f* is a rendition of the vast whirlpool discovered nearly surrounding the innermost few light-years of the Galaxy's heart. For a further description of these and other frames of this figure, consult the text.

expulsions every ten million years or so—a galactic quake of sorts.

More specifically, the curves of figure 51 depict two different theoretical models for the motion of the matter within 300 light-years of the galactic center. The solid line is computed assuming that the behavior of the obscured matter there is a mere extrapolation of the well-studied matter in the suburbs of the Galaxy near the Sun. As denoted by the solid line, the rotation velocity of the stars, gas, and dust is predicted to decrease toward the galactic center. On this basis, the innermost matter is expected to be spinning, though sluggishly.

However, radio and infrared observations strongly imply that the gas in the core region is not spinning slowly. Instead, as the

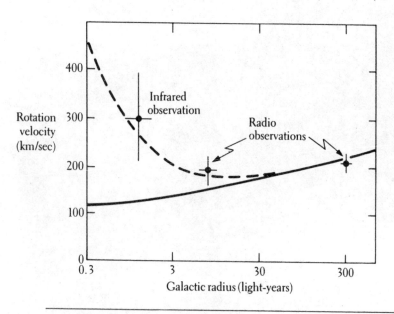

FIGURE 51 This graph plots the rotation velocity of the stars, gas, and dust near the center of our Galaxy. The solid line is a theoretical prediction of the normally expected gas motions, based on an extrapolation of the motions of matter near the Sun. The dashed curve (which is a much better fit to the observed data, shown here with their error bars) is a model of the gas motions in the presence of an immense vortex at the very heart of the Milky Way.

data points in figure 51 suggest, the observations show a dramatic rise in rotation velocity toward the center. Apparently, the heart of the Milky Way is spinning furiously; the closer to the very center, the faster the matter swirls, much like a whirlpool of water approaching a drain.

This discovery was quite a surprise to observers and theorists alike, not least because of a simple yet perplexing problem: How does such a vast galactic whirlpool maintain its structural integrity? After all, regions of rapidly rotating matter produce strong outward (centrifugal) forces tending to push the gas away, much like those that cast mud from the edge of a spinning bicycle wheel. Unless some other force pulls back on the galactic-center whirlpool, its gas should be flung into the outer parts of the Galaxy. How can such a huge vortex of matter remain intact without breaking apart and dispersing its contents? After a long list of possibilities has been eliminated, gravity is presumed to be the only viable answer.

In a simplified model capable of accounting for most of the bulk features observed in the galactic center, a (hundred square light-year) region of hot, thin, ionized gas surrounds a much smaller core of hotter, denser gas. This inner core, around which the observed gas swirls, is thought to contain—and this is the punch line—a tremendously compact object housing some five million solar masses, all packed into a region hardly larger than our Solar System. Why so much mass? Because this is the amount needed for gravity to keep the whirlpool of gas from dispersing. Such a model permits (in fact, requires, according to Kepler's laws) an increase in gas velocity toward the galactic center, as is shown by the dashed curve in figure 51, in reasonably good agreement with the observed data.

Though the details are understandably controversial at this time, the emerging consensus seems to be that a supermassive, ultracompact "something" resides in the very nucleus of the Milky Way Galaxy. And since its mass is much too large to be

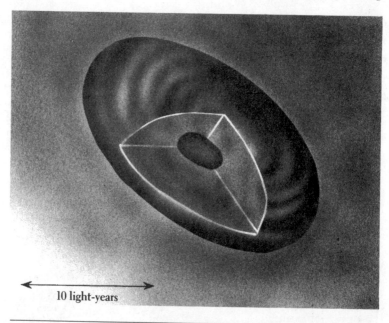

10 light-years

FIGURE 52 This is an artist's conception of a model that can generally explain most of the radio and infrared observations of the center of our Galaxy. A swirling halo is cut away to reveal a core that in turn would contain a black hole no larger, on this scale, than the period at the end of this sentence.

mistaken for any sort of ordinary star or (neutron) star remnant, and since it cannot easily be an anomalously rich star cluster, the hub of our Galaxy appears to be a huge black hole in space.

Our knowledge of the galactic center is admittedly sketchy. Astronomers are still learning to grope in the dark, literally, to decipher the clues hidden in invisible radiation. In particular, we are only beginning to appreciate the full magnitude of this entirely novel realm deep in the heart of the Milky Way. In some respects, our research should not yet even be judged as mature science. Rather, it resembles exploration—but absolutely fascinating exploration enabling us to return from our telescopes with tales of grand monuments at the core of our galactic system.

Despite our Galaxy's troublesome core, these and other issues plague our understanding of the central regions of some other galaxies. This is especially true among the minority of objects known as active galaxies, which display great upheavals. Compared with the really violent galaxies, the center of the Milky Way is quite peaceful.

Recent observations suggest that supermassive objects also lurk in or near the cores of a few other nearby galaxies. The evidence here largely resembles that for our own Galaxy; gas and stars in the innermost regions of several active galaxies are observed to be rapidly whirling. For example, observations of the M87 object shown in figure 53 imply an extremely compact source of even greater mass than exists in our normal Galaxy. In fact, several *billion* solar masses are inferred to reside within a region not much larger than the putative black hole in the Milky Way. Perhaps these central whirlpools are remnants of the turbulent eddies that helped form the galaxies in an earlier epoch.

We may well come to find that the midsection of every galaxy is inhabited by a supermassive black hole, an idea now championed by many leading astrophysicists. Normal galaxies such as our own probably house relatively small black holes having "only" millions of solar masses. More-active galaxies such as M87 might have larger holes, perhaps on the order of billions of solar masses. Of considerable import for one of the most formidable problems in all of science, the great energetics and explosiveness of the active galaxies could naturally arise from matter falling into the clutches of such supermassive black holes. After all, as we noted in chapter 17, in the process of consuming matter, black holes accelerate and heat the gravitationally infalling gas, causing it to radiate prodigious amounts of energy before entering oblivion.

It is even conceivable that the most energetic objects in the Universe—the innocuous-looking but absolutely powerhouse quasars that populate its far reaches—are ruled by hypermassive black holes that regularly consume whole stars. In fact, the energy released by a typical quasar is equivalent to the complete conversion into energy of an amount of matter equal to that in our Sun *every year.* Furthermore, quasar radiation often fluctuates from week to week, sometimes from day to day. Since cause-and-effect arguments demand that no light source flicker more quickly than

a

FIGURE 53 The ball-like M87 active galaxy lies at the center of the rich Virgo Cluster of galaxies, some 50 million light-years away, in the direction of the constellation of the same name. This series of illustrations shows *(a)* a long optical exposure of M87's halo and embedded central region, *(b)* a short optical exposure of its core and an intriguing jet of matter, and *(c)* an X-ray image highlighting the region's most active areas (including a slight asymmetry where the jet protrudes). The visual blob in frame *a* spans nearly 200,000 light-years, while the jet measures about 5,000 light-years in length and is moving away from M87's core at some 25,000 kilometers/second. The culprit presumed responsible for the intense X rays, and probably the only one that could possibly eject such a vast jet of matter, is a hypermassive black hole thousands of times larger than that thought lurking in the core of our Milky Way.

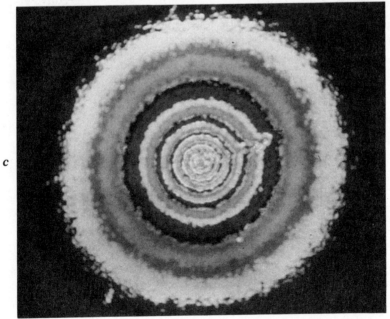

radiation can cross it (lest the observed flickering be blurred and incoherent), the huge energies of the quasars must arise from regions tiny by cosmic standards; in other words, the whole of the quasar must change at about the same time to preserve the coherency of its emitted light. How small must the quasars' central engine typically be? Hardly much larger than a few of our Solar Systems. Thus, our observations imply that quasars emit vast quantities of energy (often equivalent to about a thousand Milky Ways) while simultaneously being rather small in size. Therein lies perhaps the foremost dilemma in all of astrophysics: quasars' large energies yet small dimensions seem incompatible, especially when we realize that their radiation is often launched at all wavelengths, from those of radio waves to those of X rays. Although we currently have little observational evidence to support the idea directly, black holes in quasars would be even more massive than the billion-solar-mass objects presumed to exist in the active galaxies. Their spacetime curvature would accordingly be phenomenal, their physical properties nearly beyond human comprehension.

Clearly, an understanding of the titanic galaxies lies buried deep in their cores. We can only await future explorers who will discover, unravel, and share their secrets with us.

Unless astronomers can find direct, or very compelling indirect, evidence for the existence of black holes, neither of which we currently have, the whole concept of black holes may well turn out to be a figment of the human imagination—another case of mathematics gone awry without the check and balance of tested physical law. The nature of matter, energy, space, and time deep down inside event horizons might be no more significant than a challenging and amusing academic problem devoid of reality.

On the other hand, the Universe did emerge from what seems to have been a naked singularity (having no event horizon) some fifteen billion years ago. Black-hole singularities could well be the keys needed to unlock an understanding of the creation state from which the Universe arose. By theoretically studying the nature of black holes, and especially by experimentally seeking their existence and physical properties, we may someday be in a better position to appreciate *the* most fundamental issue of all—the origin of the Universe itself.

Epilogue
Beyond Relativity

If, then, it is true that the axiomatic basis of theoretical physics cannot be extracted from experience but must be freely invented, can we ever hope to find the right way? Nay, more, has this right way any existence outside our illusions? Can we hope to be guided safely by experience at all when there exist theories (such as classical mechanics) which to a large extent do justice to experience, without getting to the root of the matter? I answer without hesitation that there is, in my opinion, a right way, and that we are capable of finding it. Our experience hitherto justifies us in believing that nature is the realization of the simplest conceivable mathematical ideas. I am convinced that we can discover by means of purely mathematical constructions the concepts and the laws connecting them with each other, which furnish the key to the understanding of natural phenomena. Experience may suggest the appropriate mathematical concepts, but they most certainly cannot be deduced from it. . . . In a certain sense, therefore, I hold it true that pure thought can grasp reality, as the ancients dreamed.

—A. Einstein, *Ideas and Opinions*

Einstein and his colleagues in the autumn

THE SPECIAL AND GENERAL THEORIES of relativity have probably been tested more than any other theory ever proposed. This given set of ideas has serenely withstood, like no other in the history of science, a continual effort to prove them wrong, on paper and in experiments. The observational data, in particular, continue to accord well with the predictions, however baffling, of Albert Einstein. Accordingly, in the absence of evidence to the contrary, we must accept the bizarre, counterintuitive consequences of relativity theory as correct representations of physical reality. All things considered, scientists now regard relativity theory much as a defendant is viewed in the legal convention— innocent until proven guilty.

Of course, we cannot be absolutely certain that relativity theory is correct. No one can prove the complete validity of any idea. To be sure, we can never know whether a theory is precisely correct; we can know only that it is not incorrect. If a theory is proven wrong, we discard it; if it hasn't been proven wrong, we continue to test it. Thus, modern science utilizes the scientific method, again and again, to probe the accuracy of relativity under a wide variety of conditions.

Recognize that Einstein's theory has thus far been tested principally within the Solar System. Deviations from Newtonian theory are minute in our local environment; after all, no abnormally massive object resides in our own planetary system, and surely none moves at speeds close to that of light. Consequently, the weird effects of relativity are expected to be rather small in our cosmic neighborhood. So, although we have not even a hint that relativity is incorrect or incomplete, some theorists maintain that it has yet to be tested over a sufficiently broad range of conditions (but studies of gravitational lensing and binary pulsars, discussed at the end of chapter 7, hold much promise).

Similar concerns apply to relativity theory in the microscopic domain. To be sure, no one has yet proposed a convincing theory of gravitation that incorporates quantum principles. Even so, physicists conjecture that subatomic effects must become important whenever the radius of curvature of spacetime becomes less than 10^{-33} centimeter (the Planck length). On such terribly minute scales, relativity theory, which does not incorporate the Heisenberg principle, no longer seems to describe nature adequately.

Several alternative theories of gravity have been advanced in the twentieth century. Like those of Newton and Einstein, these hypotheses stress the effects resulting from rapid motions and massive objects. Significantly, all these alternatives are mathematically more complex and conceptually less simple than Einstein's version. And most of them have been experimentally proved incorrect, because their predictions disagree with one or more of the latest observational tests described earlier in this book. Some rival theories nevertheless remain tenable, largely because they and relativity predict very similar phenomena— behavior that cannot now be observationally distinguished with state-of-the-art equipment. Any one of these alternative proposals is conceivably closer to the truth than relativity, but it's likely to be some time before observations become precise enough for us to supplant Einstein's monumental achievement.

At least for the present, then, Einstein's theory remains the best known way to describe nature's behavior in the cosmos generally—a cosmos of massive objects, high velocities, significant curvature, and subtle departures from Newtonian common sense. This statement admittedly embodies some risk, for the effects of relativity throughout the whole Universe are nearly a million times greater than in our terrestrial niche of the Solar System. But

Selected Further Readings

WORKS ROUGHLY AT THE LEVEL OF THIS BOOK

Barker, Peter, and Cecil G. Shugart, eds. *After Einstein.* Memphis: Memphis State Univ. Press, 1981.
Barnett, Lincoln K. *The Universe and Dr. Einstein.* New York: Mentor, 1948.
Bernstein, Jeremy. *Einstein.* New York: Viking, 1973.
Calder, Nigel. *Einstein's Universe.* New York: Viking, 1979.
Clark, Ronald W. *Einstein: The Life and Times.* New York: World, 1971.
Field, George B., and Eric J. Chaisson, *The Invisible Universe: Probing the Frontiers of Astrophysics.* Boston: Birkhauser, 1985.
Gardner, Martin. *The Relativity Explosion.* New York: Random House, 1976.
Harrison, Edward R. *Cosmology: The Science of the Universe.* New York: Cambridge Univ. Press, 1981.

MORE ADVANCED WORKS

French, A. P. ed. *Einstein: A Centenary Volume.* Cambridge: Harvard Univ. Press, 1979.
Miller, Arthur I. *Albert Einstein's Special Theory of Relativity: Emergence (1905) and Early Interpretation (1905–11).* Boston: Birkhauser, 1981.

Misner, Charles W., Kip Thorne, and John A. Wheeler. *Gravitation.* San Francisco: Freeman, 1973.

North, John D. *The Measure of the Universe.* Oxford: Clarendon Press, 1965.

Pais, Abraham. *'Subtle Is the Lord . . .': The Science and Life of Albert Einstein.* New York: Oxford Univ. Press, 1982.

Singh, Jagjit *Great Ideas and Theories of Modern Cosmology,* 2d ed. rev. New York: Dover, 1970.

Weinberg, Steven. *Gravitation and Cosmology: Principles and Applications of the General Theory of Relativity.* New York: Wiley, 1972.

Woolf, Harry, ed. *Some Strangeness in the Proportion.* Reading, Mass.: Addison-Wesley, 1980.

SOME WORKS BY EINSTEIN

The Meaning of Relativity. 5th ed. Princeton: Princeton Univ. Press, 1956.

Relativity: The Special and General Theory. New York: Crown, 1961.

The World as I See It. Translated by Alan Harris. New York: Philosophical Library, 1949.

Out of My Later Years. New York: Philosophical Library, 1950.

The Evolution of Physics, with Leopold Infeld. New York: Simon & Schuster, 1954.

Ideas and Opinions. Based on *Mein Weltbild,* ed. Carl Seelig, and other sources. Translated and revised by Sonja Bargmann. New York: Crown, 1954.

Index

Index

Index

ABOUT THE AUTHOR

ERIC J. CHAISSON *has published approximately a hundred scientific articles, most of them in the professional journals. He has also written several books, including* The Life Era *and* Cosmic Dawn, *which won the Phi Beta Kappa Prize, the American Institute of Physics Award, and an American Book Award Nomination for distinguished science writing. Trained initially in condensed-matter physics, Chaisson received his doctorate in astrophysics from Harvard University, where he spent a decade as a member of the Faculty of Arts and Sciences. While at the Harvard-Smithsonian Center for Astrophysics, he won fellowships from the National Academy of Sciences and the Sloan Foundation as well as Harvard's Bok and Smith Prizes. Dr. Chaisson is currently a senior scientist and division head at the Space Telescope Science Institute on the Johns Hopkins campus. He lives with his wife and two children in Annapolis.*

ABOUT THE ILLUSTRATOR

LOLA JUDITH CHAISSON *(née Eachus) majored in astronomy at Wellesley College and worked as an astronomer and illustrator at the Harvard-Smithsonian Center for Astrophysics before becoming a free-lance illustrator and mother of two.*